MORPHOLOGY AND BIONOMICS OF DORYLAIMS (NEMATODA: DORYLAIMIDA)

Top left: Anterior region of *Nevadanema nevadense* from Spain. Top centre: Vagina of *Labronema* sp. from Vietnam. Bottom left: In face view of the lip region of *Aporcelaimellus communis* from Spain. Bottom centre: Spicule of *Sectonema* sp. from Vietnam. Right: Anterior region of *Pachydorylaimus holovachovi* from Costa Rica.

MORPHOLOGY AND BIONOMICS OF DORYLAIMS (NEMATODA: DORYLAIMIDA)

Reyes Peña-Santiago

David J. Hunt and Roland N. Perry (Series Editors)

NEMATOLOGY MONOGRAPHS AND PERSPECTIVES
VOLUME 13

TUTA SUB AEGIDE PALLAS · 1683 ·

BRILL
LEIDEN-BOSTON
2021

This book is printed on acid-free paper.

Library of Congress Cataloging-in-Publication Data

Names: Peña-Santiago, R. (Reyes), author.
Title: Bionomics of dorylaims (nematoda, dorylaimida) / Reyes Peña Santiago.
Description: Leiden; Boston: Brill, [2021] | Series: Nematology monographs and perspectives, 1573-5869; volume 13 | Includes bibliographical references and index.
Identifiers: LCCN 2020052808 (print) | LCCN 2020052809 (ebook) | ISBN 9789004439993 (hardback) | ISBN 9789004447851 (ebook)
Subjects: LCSH: Dorylaimida. | Dorylaimida–Morphology. | Dorylaimida–Behavior. | Dorylaimida–Ecology.
Classification: LCC QL391.N4 P43 2021 (print) | LCC QL391.N4 (ebook) | DDC 592/.57–dc23
LC record available at https://lccn.loc.gov/2020052808
LC ebook record available at https://lccn.loc.gov/2020052809

ISBN: 978 90 04 43999 3
E-ISBN: 978 90 04 44785 1

A mi esposa, Mari Carmen, a mi hija, María Victoria, y a mi hijo, Juan Enrique, por el tiempo que he dedicado a los doriláimidos y que no he pasado con ellos.

Contents

Nematology Monographs & Perspectives, 2021, Vol. 13, xiii-xiv

Foreword

The dorylaims are one of the most diverse and ecologically important nematode groups of soil and freshwater nematodes and have been the subject of much research. Jairajpuri & Ahmad (1992) recognised about 200 genera with 1700 species in the Dorylaimida. According to Andrássy (2009) the number of valid Dorylaimida genera had increased to 263 with over 2600 valid species, a substantial rate of increase. Not surprisngly, the group continues to accrue new species and genera from many different habitats and countries.

Of the many important publications on this group, mention must be made of the monographs by Thorne (1939), Jairajpuri & Ahmad (1992) and Andrássy (2009), these volumes usually focusing rather more on taxonomy and systematics than morphology, bionomics and biogeography. The present volume, prepared by Prof. Reyes Peña-Santiago, one of the foremost and most experienced experts in the group, redresses this balance, although also including a very useful, updated and comprehensive, systematic scheme for the group.

Much of the more recent interest in dorylaims has been related to their potential role as ecological markers or bio-indicators in soil, an area where their highly variable feeding habits and enormous variation in body form and size makes them particularly useful in assessing soil status or 'health' and other environmental variables. The fact that many species show a high sensitivity to heavy metals and various soil contaminants adds to their portfolio of attributes in this field.

Dorylaims display a great variety of lifestyles and are abundant, probably the most abundant, components in both soil and freshwater environments. Some are fungal feeders, whilst others are algivorous, omnivorous, predatory or obligate plant-parasites. As a result of this diverse feeding strategy, their morphology is, although clearly derived from a basic pattern, also highly variable, particularly in relation to the structure of the lip region and associated structures, such as the odontostyle and odontophore, but also in the highly variable pharynx, form of the genital tract in both sexes and body form, such as overall size and tail shape. Although many of the 'classical' genera and higher taxa are likely polyphyletic, molecular sequencing is beginning to shed light on the complex, not to mention confusing, relationships that often exist between morphological concepts and phylogeny. The fact remains, however, that many dorylaims lack molecular data and morphology remains a crucial diagnostic parameter. On a morphological level the dorylaims present a number of challenges to the tyro nematologist or general soil biologist. It is thus vitally important that researchers intending to study the group, for whatever purpose, are familiar with the often unique morphology of these fascinating creatures.

The focus of this volume by Reyes Peña-Santiago, therefore, is to present an account of the diverse morphology and bionomics within the group and thereby serve as a comprehensive introduction and handbook to the dorylaims. The book, of necessity, is well illustrated by carefully selected figures and microphotographs, many of which stem from the author's own research conducted over several decades. Students of the group will find a wealth of information summarised within these pages and it should be essential reading for any aspiring dorylaimophile.

David J. HUNT
Emeritus Fellow, CABI
Egham, Surrey, TW20 9TY
UK

References

ANDRÁSSY, I. (2009). *Free-living nematodes of Hungary, III (Nematoda errantia)*. In: *Pedozoologica Hungarica, 5* (Series Editors: Csuzdi, C. & Mahunka, S.). Budapest, Hungary, Hungarian Natural History Museum and Systematic Zoology Research Group of the Hungarian Academy of Sciences.

JAIRAJPURI, M.S. & AHMAD, W. (1992). *Dorylaimida. Free-living, predaceous and plant-parasitic nematodes*. Leiden, The Netherlands, Brill.

THORNE, G. (1939). A monograph of the nematodes of the superfamily Dorylaimoidea. *Capita Zoologica* 8, 1-261.

Nematology Monographs & Perspectives, 2021, Vol. 13, xv-xvii

Preface

With more than 3300 valid species, the members of the order Dorylaimida, often termed dorylaims, probably constitute one of the most diverse taxa in the phylum Nematoda. They are continental forms, very abundant and frequent in both soils and freshwater sediments around the world, and totally absent in marine habitats with the exception of a few species occasionally recorded in estuarine environments. Many aspects of their biology remain poorly explored, but most of their taxa are regarded as free-living forms with omnivorous and/or predacious feeding habits, although the representatives of the family Longidoridae are reputed plant parasites and the occurrence of dorylaims in the body cavity of other animals is extremely occasional and rather anecdotal. In general, and under an ecological perspective, dorylaims are considered to be persisters and K-strategists due to their long generation time, low reproduction rates, low metabolic activity and slow movement. Their role in the habitats they dwell in has not yet been sufficiently elucidated. Nevertheless, their contribution to nematode biomass in soils and sediments is certainly notable as they show comparatively large-sized bodies, usually 1-3 mm long and sometimes surpassing 4 or 5 mm. Because of their high diversity and abundance, as well as their ecological traits, dorylaims have become good bioindicators of environmental quality (health).

Morphologically, dorylaims are an easily recognisable taxon. They form a well-defined, monophyletic group whose members share a series of apomorphic, even autapomorphic, conditions: stoma armed with a protrusible structure (either a mural tooth or an axial odontostyle), two-sectioned (bottle-like) pharynx, prerectum (specialised posterior portion of intestine, unique within Nematoda), one pair of male adcloacal genital papillae (exclusive to the group), and caudal glands absent. Nonetheless, leaving aside this basic and distinguishable pattern, dorylaims show an intricate morphological diversity mainly affecting their lip region, the nature of the stomatal protrusible structure, pharynx, female and male genital systems, and caudal region. The interpretation and the understanding of the myriad of observable traits in these nematodes are fundamental for the study of their taxonomy and phylogeny. Luckily, the literature available on the morphology of dorylaims displays an excellent quality, even if, unfortunately, it is scattered in many papers and monographs and has not been updated for three decades since the 1992 publication of the essential *Dorylaimida* by Jairajpuri and Ahmad.

The aim of this contribution is to show the reader what dorylaims are like and how they live. Its content is divided into 14 chapters. The book starts with a synthetic concept of the group, including their definition and summarised ideas about their biology, distribution and diversity. The morphology of the group occupies most of the monograph, where relevant traits for their characterisation, identification and classification are described and discussed. This part consists of nine chapters dealing with general aspects (size, shape, habitus, colour), body wall and pseudocoel, lip region and amphids, stoma and feeding apparatus, digestive tract, female genital system, male genital system, nervous systems and receptors, and caudal region. The second part is devoted to key aspects of the biology of the group, namely feeding habits, reproduction and development, ecology and biogeography, and diversity. The text of each chapter is supplemented by illustrations and a list of references. The former, either line drawings and/or (LM, SEM) microphotographs, are useful to understand the literal description of the many traits and processes. The latter provides an additional source of information concerning each item.

This book is conceived as a tool for young nematologists who wish to start out on a study of the Dorylaimida, but is also aimed at more experienced scientists who do not have too much experience with dorylaims and who, for any reason, need to work with them. Its content is based on my personal notes and presentations, which were designed to give several courses on the diversity and identification of free-living Dorylaimia at the Universities of Ghent (Belgium) and Jaén (Spain), within the framework of the European Master in Nematology that, in its different versions and periods, was conducted and led by reputed colleagues of Ghent University. My experience in teaching this Master, often aimed at students from developing countries with hardly any previous nematological background, prompted me to compile my notes and to organise them in the form of a monograph.

After a 40 year career in nematology, uninterruptedly devoted to exploring and explaining the diversity of Dorylaimida, this book contains not only part of my teaching experience but also many inputs from my research path. Nonetheless, being also a compilation of current knowledge of dorylaims, much of its content is inspired by previous works, hence, the pioneering contributions by just ten nematologists (N.A. Cobb, G. Thorne, I. Andrássy, M.R. Siddiqi, P.A.A. Loof, J. Heyns, M.S. Jairajpuri, M.T. Vinciguerra and W. Ahmad) who formed the ground plan of dorylaimid diversity and biology, and who serve as obligatory references. Still, many ideas and a good part of the line illustrations are essentially based on, when not directly extracted from, a series of unsurpassable publications by Prof. A. Coomans (and his collaborators at Ghent University), who taught us to look at, to understand and to interpret fundamental aspects of dorylaimid morphology. In this sense, this book is also a tribute to all of them.

Reyes PEÑA-SANTIAGO
Jaén (Spain), July 2020

Acknowledgements

I would like to thank the University of Jaén for the financial and technical support received through the research programmes 'PAIJUA 2017/2018 and 2019/2020: EI_RNM02_2017 and 2019' for my scientific work throughout the last four years, when this book was conceived and written. My gratitude also to María del Mar López-Castro, staff member of the Library of the Universisty of Jaén, who assisted me in accessing old and/or rare nematological publications. Most line illustrations have been taken from many authors and contributions, all of them cited, referenced in detail, and deeply appreciated. Finally, but no less important, the advice received from David Hunt and Roland Perry, in particular their editoral comments and suggestions and, very especially, their revision of the English text, have been of inestimable value and deserving of my enormous gratitude.

Nematology Monographs & Perspectives, 2021, Vol. 13, 1-6

Chapter 1

Concept

Dorylaims, the representatives of the order Dorylaimida Pearse, 1942, constitute one of the most diverse taxa in the phylum Nematoda and are a natural group whose monophyly is supported by both morphological and molecular evidence. They display a recognisable morphological pattern, characterised by a unique combination of traits, either apomorphic or (even) autapomorphic conditions. Their feeding habits have not been totally elucidated, but it is often assumed that most species are free-living, omnivorous or predatory, although it has been demonstrated that they can use other resources too. Very abundant and frequent in continental habitats, especially in pristine soils and freshwater sediments, their occurrence in marine systems is simply anecdotal. Because their abundance and diversity decrease significantly in perturbed or polluted habitats, dorylaims are often regarded as good environmental bio-indicators.

Morphological characterisation: distinctive features

A combination of five traits (Fig. 1.1) sets up an easily recognisable dorylaimid morphological pattern, which differs from those of other nematode orders (Peña-Santiago, 2014).

The stoma is always armed with a protruding structure, either a mural tooth (Fig. 1.1D) or an axial odontostyle (Fig. 1.1B) originating within the pharyngeal tissues, extremely variable in shape and size, and movable by means of protractor muscles. Similar forms of the stomatal armature are found in other nematodes, for example, tylenchs, aphelenchs (Rhabditida, Tylenchomorpha) and triplonchs (Triplonchida), but they seem to be significantly different. Thus, the typical spear of tylenchs and aphelenchs is a stomastostylet, *i.e.*, a modified stoma (Maggenti, 1981; Baldwin & Perry, 2004), whereas the characteristic onchiostyle of triplonchs is entirely lodged within, and projects into, the pharyngeal lumen (Maggenti, 1981; Decraemer, 1995). The available infor-

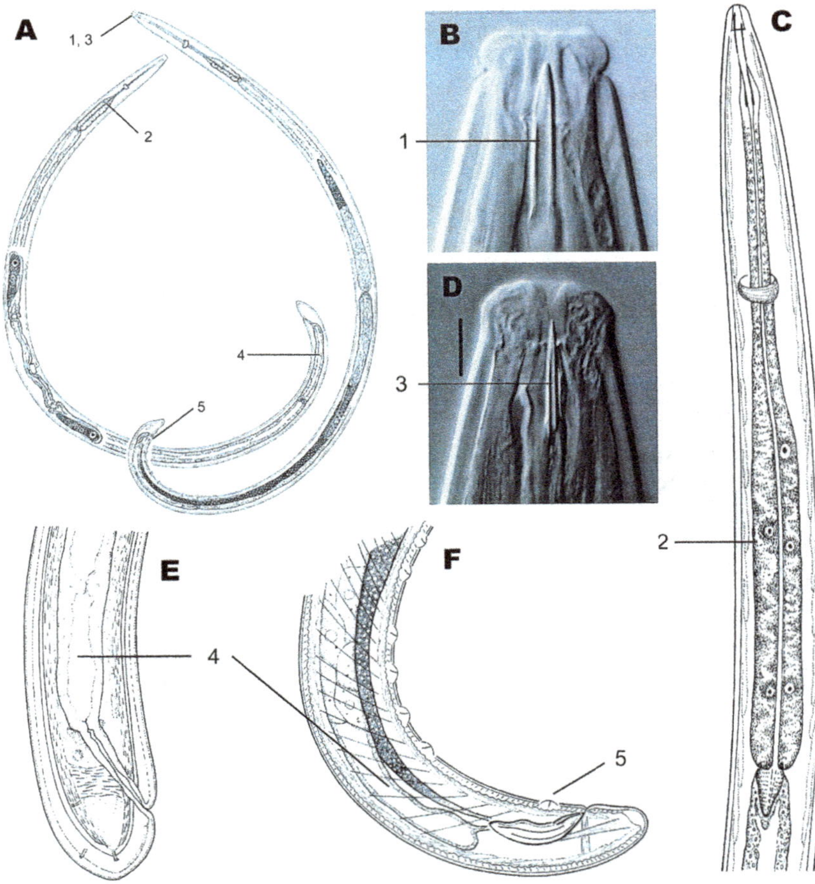

Fig. 1.1. *Diagnostic (apo- or autapomorphic) features of Dorylaimida. A: Female and male entire body; B: Anterior region of a representative of Dorylaimina, with axial odontostyle (1); C: Neck region, with bottle-like pharynx (2); D: Anterior region of a representative of Nygolaimina, with mural tooth (3); E: Posterior region of female, with prerectum (4); F: Posterior region of male, with prerectum (4) and adcloacal pair of genital papillae (5). (After Peña-Santiago (2014).* Handbook of Zoology – Gastrotricha, Cycloneuralia and Gnathifera. Volume 2. Nematoda*: p. 278, Fig. 75.1.)*

mation does not allow deep comparisons of the nature and embryological origin of these protruding structures (Maggenti, 1981; Baldwin & Perry, 2004) but, as currently conceived, the dorylaimid 'spear' (mural tooth or odontostyle) might represent an autapomorphy of the group.

The pharynx is bipartite and bottle-like (Fig. 1.1C), *i.e.*, it typically consists of two sections: an anterior, more slender region, which then enlarges into an expanded, more posterior (basal) portion. The latter contains five (one dorsal and two pairs of laterosubventral), more rarely three (lacking one of the two pairs), mono-nucleate glands with their respective outlets. This same scheme has also been reported in other nematode taxa, for example, alaims (Enoplida, Alaimina; see Choudhary *et al.*, 2010) and triplonchs (Decraemer, 2006). Pharynx morphology should therefore also be considered as an apomorphic condition in dorylaims.

The posterior region of the intestine, termed the prerectum (Fig. 1.1E_4, F_4), is especially differentiated, being set off from the anterior region of (the true) intestine by the existence of a transverse septum, and from the rectum by a narrowed section surrounded by a muscular sphincter. Such differentiation, found in all the known dorylaims and never reported in other nematode taxa, represents a conspicuous autapomorphic condition of the group.

Male dorylaims bear genital papillae (also termed supplements) along the mid-ventral line of the posterior body region. One pair of these papillae is situated immediately anterior to the cloacal aperture (Fig. 1.1F_5) and forms the pre- or adcloacal supplements, whereas the remaining supplements are single and form a ventromedian series arranged in tandem. A comparable ventromedian series of genital papillae has been reported in other nematodes, such as alaims or mononchs (Mononchida; see Ahmad & Jairajpuri, 2010), but the existence of an adcloacal pair has been described in all the bisexual dorylaimid species and is exclusive to the group. Although the functional meaning of this evolutionary novelty is not well understood, it represents another remarkable autapomorphy of the dorylaims.

Lastly, dorylaims completely lack caudal glands, which have been probably lost as a result of their adaptation to continental and, more specifically, terrestrial habitats. The absence of caudal glands should be interpreted as an apomorphic condition as it also occurs in many other nematode taxa.

Biology

Many features of dorylaimid biology remain obscure or poorly explored, but relevant advances have been achieved in the elucidation of some aspects. Dorylaims have traditionally been considered to be omnivores and/or predators (Yeates *et al.*, 1993). Nonetheless, the representatives of the specialised family Longidoridae and a few particular genera are recognised ectoparasites, feeding on vascular plants using a piercing and suction technique. Besides, observational and experimental evidence suggests that some species might be fungivorous or algivorous, and even saprophagous, consuming organic materials undergoing decomposition (Wood, 1973, 1974; Okada *et al.*, 2006).

Sexual reproduction is the only mode of reproduction known to occur in dorylaims and amphimixis is the most usual process, and, although parthenogenesis is also of common occurrence, there is no good evidence of hermaphroditism. Their postnatal development has been well documented and follows the general nematode pattern with four (exceptionally three) juvenile stages (J1-J4). In general, dorylaims are K-strategists, their life-cycle requiring a period of at least several weeks to be completed.

Distribution

Dorylaims are major components (high diversity, abundance and biomass) of the nematode communities dwelling in terrestrial soils and freshwater sediments but, unfortunately, the study of their ecology has not so far received much attention. Thus, their role in feeding webs has not been well elucidated. They are distributed worldwide in continental habitats, but their geographical ranges are still far from being well delimited. A few species, for example, *Aporcelaimellus obtusicaudatus* and *Ecumenicus monohystera*, are probably cosmopolitan, but it is certain that many others have more restricted ranges, and some genera are distributed in particular areas, as observed in *Pachydorylaimus*, an endemic genus from the Neotropical region. However, it has been demonstrated that, at least in natural (non-perturbed) habitats, dorylaimid diversity is much higher in general than that of any other nematode group. For example, a total of 610 dorylaimid species inhabit freshwater bodies around the world, more than thrice the number of Rhabditida which, with 164 species, are the second most diverse group (Abebe *et al.*, 2008).

Many environmental variables may affect the general distribution of dorylaims, their frequency and abundance, but only a few (one to three) of these many variables are usually enough to explain the distribution of individual species in some territories (Liébanas *et al.*, 2004).

Diversity

With more than 3300 valid species and more than 250 valid genera (Andrássy, 2009), Dorylaimida are probably the most diverse nematode order. The foundations of their classification were set by Gerald Thorne in a series of contributions that appeared in the 1930s (see Thorne, 1939), whereas the second half of the 20th century brought significant progress in the understanding and interpretation of their body plan, as well as extraordinary advances in the knowledge of their diversity through the discovery of hundreds of new taxa (see Jairajpuri & Ahmad, 1992). Among the outstanding landmarks of the new millennium is the integrative approach used to explore the phylogeny and systematics of the group by combining both morphological and molecular data (see De Ley & Blaxter, 2004). Nevertheless, leaving aside their two, well-supported, monophyletic suborders, the Dorylaimina and Nygolaimina, the classification of dorylaims is not totally satisfactory as the internal relationships among the main subgroups are still weakly established and the traditional system of three superfamilies (Dorylaimoidea, Belondiroidea and Tylencholaimoidea) does not match the trees derived from molecular analyses.

References

ABEBE, E., DECRAEMER, W. & DE LEY, P. (2008). Global diversity of nematodes (Nematoda) in freshwater. *Hydrobiologia* 595, 67-78. DOI: 10.1007/s10750-007-9005-5

AHMAD, W. & JAIRAJPURI, M.S. (2010). *Mononchida – the predatory soil nematodes. Nematology Monographs and Perspectives 7* (Series Editors: Hunt, D.J. & Perry, R.N.). Leiden, The Netherlands, Brill.

ANDRÁSSY, I. (2009). *Free-living nematodes of Hungary, III (Nematoda errantia)*. In: *Pedozoologica Hungarica 5* (Series Editors: Csuzdi, C. & Mahunka, S.). Budapest, Hungary, Hungarian Natural History Museum and Systematic Zoology Research Group of the Hungarian Academy of Sciences.

BALDWIN, J.G. & PERRY, R.N. (2004). Nematode morphology, sensory structure and function. In: Chen, Z.X., Chen, S.Y. & Dickson, D.W. (Eds). *Nematology: advances and perspectives, Volume I – Nematode morphology, physiology and ecology*. Wallingford, UK, CAB International, pp. 175-257.

CHOUDHARY, M., AHMAD, W. & JAIRAJPURI, M.S. (2010). *Alaimina. Free-living soil inhabiting nematodes*. Aligarh, India, Aligarh Muslim University Press.

DECRAEMER, W. (1995). *The family Trichodoridae: stubby root and virus vector nematodes*. Dordrecht, The Netherlands, Kluwer Academic Publishers.

DECRAEMER, W. (2006). Order Triplonchida. In: Eyualem-Abebe, Andrássy, I. & Traunspurger, W. (Eds). *Freshwater nematodes: ecology and taxonomy*. Wallingford, UK, CAB International, pp. 293-325.

DE LEY, P. & BLAXTER, M. (2004). A new system for Nematoda: combining morphological characters with molecular trees, and translating clades into ranks and taxa. In: Cook, R. & Hunt, D.J. (Eds). *Proceedings of the Fourth International Congress of Nematology, 8-13 June 2002, Tenerife, Spain. Nematology Monographs and Perspectives 2* (Series Editors: Cook, R. & Hunt, D.J.). Leiden, The Netherlands, Brill, pp. 633-653.

JAIRAJPURI, M.S. & AHMAD, W. (1992). *Dorylaimida. Free-living, predaceous and plant-parasitic nematodes*. Leiden, The Netherlands, Brill.

LIÉBANAS, G., GUERRERO, P., MARTÍN-GARCÍA, J.M. & PEÑA-SANTIAGO, R. (2004). Spatial distribution of dorylaimid and mononchid nematodes from southeast Iberian Peninsula: environmental characterization of chorotypes. *Journal of Nematology* 36, 114-122.

MAGGENTI, A. (1981). *General nematology*. New York, NY, USA, Springer-Verlag Inc.

OKADA, H., HARADA, H., TSUKIBOSHI, T. & ARAKI, M. (2006). Characteristics of *Tylencholaimus parvus* (Nematoda: Dorylaimida) as a fungivorus nematode. *Nematology* 7(2005), 843-849. DOI: 10.1163/156854105776186424

PEÑA-SANTIAGO, R. (2014). Order Dorylaimida Pearse, 1942. In: Schmidt-Rhaesa, A. (Ed.). *Handbook of zoology – Gastrotricha, Cycloneuralia and Gnathifera. Volume 2. Nematoda*. Berlin, Germany, De Gruyter, pp. 277-297.

THORNE, G. (1939). A monograph of the nematodes of the superfamily Dorylaimoidea. *Capita Zoologica* 8, 1-261.

WOOD, F.H. (1973). Nematode feeding relationships. Feeding relationships of soil-dwelling nematodes. *Soil Biology and Biochemistry* 5, 593-601. DOI: 10.1016/0038-0717(73)90049-7

WOOD, F.H. (1974). Biology of *Aporcelaimellus* sp. (Nematoda: Aporcelaimidae). *Nematologica* 19(1973), 528-537. DOI: 10.1163/187529273X00547

Nematology Monographs & Perspectives, 2021, Vol. 13, 7-12

Chapter 2

General aspect

Size

Dorylaims are comparatively large nematodes. Their length usually ranges from 1 to 3 mm, but often exceeds these limits, with some species less than 0.5 mm and a few about 10 mm long. The genus *Tylencholaimus* probably contains some of the smallest forms, for example, *T. pusillus* from India and *T. micronanus* from New Zealand, which are 0.30-0.40 (Loof & Jairajpuri, 1968) and 0.37-0.42 mm long (Yeates, 1979), respectively. Among the free-living forms, the largest species are found within the genera *Dorylaimus* and *Tubixaba*, with *D. macrosoma* from the Iberian Peninsula, and *T. tuxaua* from Brazil, reaching a body length of 9.1 and 10.0 mm, respectively. Plant-parasitic species may reach and exceed these values, as occurs with *Paralongidorus maximus*, its body length ranging from 7.60 to 12.35 mm in German populations (Sturhan, 1963). Many species belonging to *Eudorylaimus*, *Aporcelaimellus* or *Mesodorylaimus*, three of the most diverse and wide-spread genera in both soil and freshwater habitats, seldom fall under 1 mm or surpass 3 mm long. For descriptive purposes, a literal reference to body size (length) may be done following the guide in Table 2.1.

Table 2.1. *Relationship between body length and descriptive terminology.*

Length range (mm)	Literal description
<0.5	very small
0.5-1.0	small
1.0-3.0	medium
3.0-5.0	large
>5.0	very large

Table 2.2. *Relationship between Demanian ratio a values and descriptive terminology.*

Ratio a value	Literal description
<20	stout (obese)
20-30	moderately slender
30-40	slender
>40	very slender

Body diameter also displays broad variation. Its maximum, usually reached at the mid-body region, ranges from less than 20 μm in some *Discolaimium* species to more than 200 μm in members of the genus *Dorylaimus* and the family Aporcelaimidae. The ratio 'body length divided by body diameter', *i.e.*, the Demanian ratio 'a', provides a good approximation to the general dimension, in particular the slenderness or thinness, of nematodes. In dorylaims, ratio a very often varies from 20 to 30, rarely surpassing these limits, especially the lower one. With due caution, and for descriptive purposes, the guide in Table 2.2 can be used to refer literally to this criterion.

Shape

The body of dorylaims is nearly cylindrical throughout its length and more or less tapering gradually towards both extremities. It narrows conspicuously from the neck base to the anterior end (lip region), the width (diameter) of the latter becoming about one-third of that of the former (Fig. 2.1B). Occasionally, some species, for example, the representatives of the genus *Paraxonchium*, display a more conspicuously tapering neck, in which the lip region becomes very narrow (Fig. 2.1A), being one-sixth of the body diameter at the neck base or even less. Conversely, members of the genus *Discolaimus* and a few other genera are characterised by their expanded, disc-like lip region (Fig. 2.1C). The body nearly always tapers to the posterior region, in particular from the female anus and the male cloacal aperture to the tail tip. This is easily seen in those forms that bear a long filiform (Fig. 2.1D) or conical (Fig. 2.1E) caudal region. Nevertheless, the tapering is less discernible in species displaying a short and rounded (Fig. 2.1G) or cylindrical (Fig. 2.1F) caudal region. Very exceptionally, the posterior end may even appear visibly thickened or clavate (Fig. 2.1H).

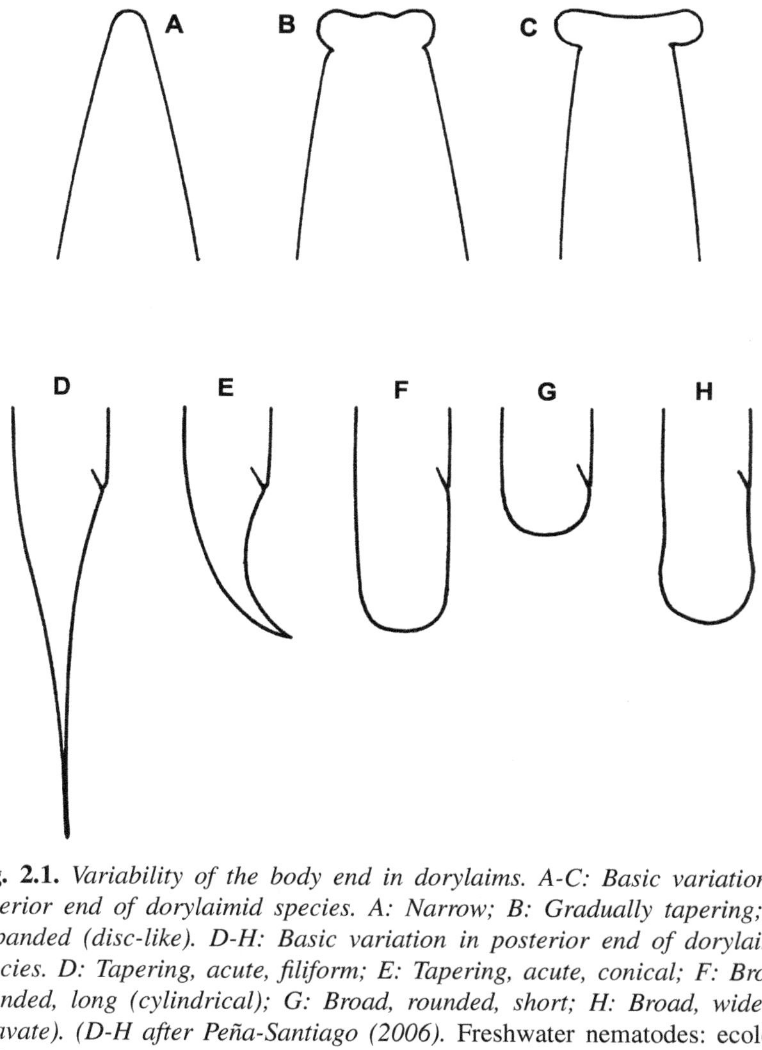

Fig. 2.1. *Variability of the body end in dorylaims. A-C: Basic variation in anterior end of dorylaimid species. A: Narrow; B: Gradually tapering; C: Expanded (disc-like). D-H: Basic variation in posterior end of dorylaimid species. D: Tapering, acute, filiform; E: Tapering, acute, conical; F: Broad, rounded, long (cylindrical); G: Broad, rounded, short; H: Broad, widened (clavate). (D-H after Peña-Santiago (2006).* Freshwater nematodes: ecology and taxonomy*: p. 338, Fig. 14.5B-F.)*

Habitus

Upon killing and fixation, especially if the animals were previously relaxed, the body posture (habitus) of dorylaims is rather variable. It may become totally straight to strongly curved ventrad or G-shaped (Figs 2.2, 2.3). A straight body is very unusual, but is often found in representatives

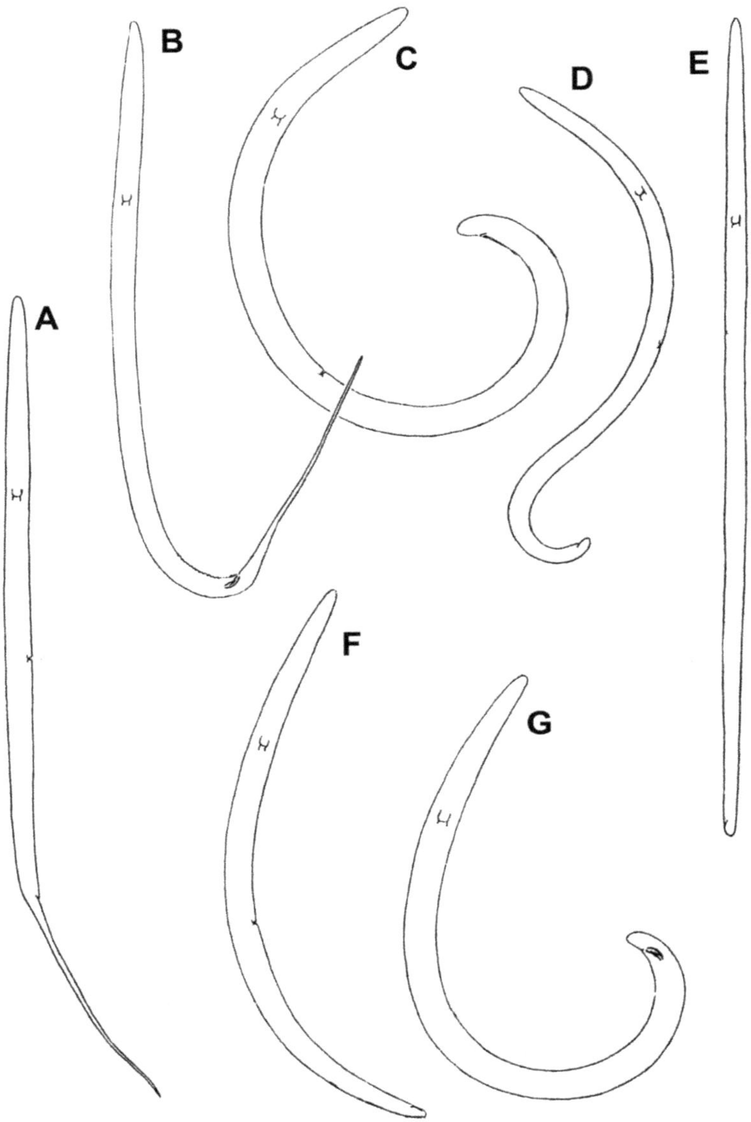

Fig. 2.2. *Variability of the body posture (habitus) in dorylaims. A:* Prodorylaimus *sp., female; B:* Prodorylaimus *sp., male; C:* Xiphinema *sp., female; D:* Solididens *sp., female; E:* Belondira *sp.; F:* Enchodelus *sp., female; G:* Enchodelus *sp., male. (After Jairajpuri & Ahmad (1992).* Dorylaimida*: p. 2, Fig. 1.)*

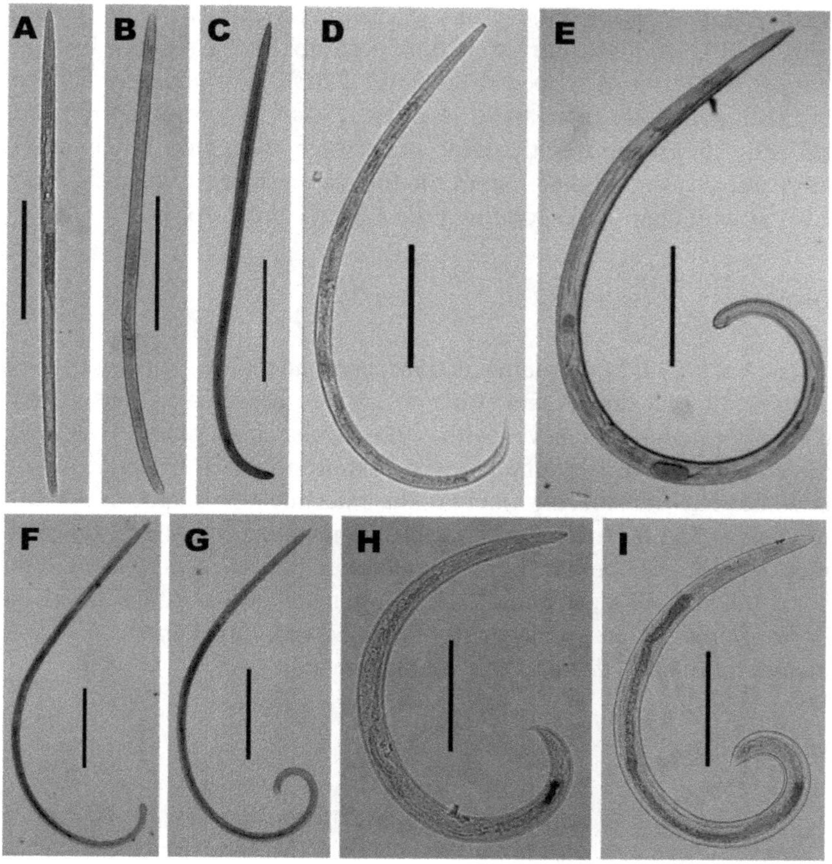

Fig. 2.3. *Variability of the body posture (habitus) in dorylaims (LM). A: Straight,* Belondira *sp. from Vietnam; B: Weakly curved ventrad,* Adenolaimus tropicus *from Iran; C: J-shaped,* Syncheilaxonchium nairi *from Iran; D: J-shaped,* Oriverutus vanisti *from Japan; E: G-shaped,* Sectonema *sp. from Vietnam; F: J-shaped, female* Axonchoides smokyensis *from the USA; G: G-shaped, male* Axonchoides smokyensis *from the USA; H: G-shaped, female* Oriverutus vinciguerrae *from Iran; I: G-shaped, male* Oriverutus vinciguerrae *from Iran. (Scale bars: A, B, D, H, I = 200 μm; C, E-G = 500 μm.)*

of some genera, for example *Belondira* (Figs 2.2E; 2.3A). A sigmoid, that is an S-shaped, body (Fig. 2.2D) occasionally occurs in some nygolaims (Nygolaimina). Leaving aside these infrequent variations, the habitus is almost always more or less curved ventrad: weakly arcuate (Fig. 2.3B), regularly curved ventrad adopting an open C-shape (Fig. 2.2F), or more curved in the posterior body region, J-shaped (Fig. 2.3C, D) or G-shaped (Figs 2.2C; 2.3E). As frequently happens in other nematode taxa, there is often a marked sexual dimorphism affecting this feature as the male posterior body region is commonly more curved ventrad than that of the female (compare Figs 2.2A, F and 2.3F, H with their corresponding Figs 2.2B, G and 2.3G, I).

Colour

The body wall of dorylaims is transparent and allows the observation of their internal organs and structures, which are usually consistent in their appearance, but never with a distinguishable colour. Thus, they are colourless animals. The pharynx, mainly made of muscular tissue, is somewhat translucent, whereas the intestine often appears slightly brownish. The material found within the intestine may have, however, more appreciable colour. For example, members of *Heterodorus* often bear greenish food, probably algae, in their gut, and specimens of *Aporcelaimellus* are occasionally found containing darker (blackish) material, probably decomposing organic matter.

References

JAIRAJPURI, M.S. & AHMAD, W. (1992). *Dorylaimida. Free-living, predaceous and plant-parasitic nematodes*. Leiden, The Netherlands, Brill.

LOOF, P.A.A. & JAIRAJPURI, M.S. (1968). Taxonomic studies on the genus *Tylencholaimus* de Man, 1876 (Dorylaimoidea) with a key to the species. *Nematologica* 14, 317-350. DOI: 10.1163/187529268X00011

PEÑA-SANTIAGO, R. (2006). Dorylaimida Part I: Superfamilies Belondiroidea, Nygolaimoidea and Tylencholaimoidea. In: Eyualem-Abebe, Traunspurger, W. & Andrássy. I. (Eds). *Freshwater nematodes: ecology and taxonomy*. Wallingford, UK, CAB International, 326-391.

STURHAN, D. (1963). Beitrag zur Systematik der Gattung *Longidorus*. *Nematologica* 9, 131-142. DOI: 10.1163/187529263X00241

YEATES, G.W. (1979). Nine new Dorylaimida (Nematoda) from the New Zealand region. *Nematologica* 25, 419-438. DOI: 10.1163/187529279X00587

Nematology Monographs & Perspectives, 2021, Vol. 13, 13-25

Chapter 3

Body wall and pseudocoel

Dorylaims are no exception to the nematode morphological pattern. Thus, their body wall consists of three basic elements (Lippens, 1975) (Fig. 3.1A): outer cuticle, intermediate epidermis (also termed the hypodermis), and inner longitudinal somatic musculature (internal). The body wall envelops the body cavity or pseudocoel.

Cuticle

STRUCTURE (LAYERS)

The cuticle of dorylaims is a complex layered component of the body wall. As revealed by transmission electron microscopy (TEM), it is made of seven layers (Roggen *et al.*, 1967; Aboul-Eid, 1969; Lippens *et al.*, 1974; Lippens, 1975; Heyns & Jacobs, 1989), which can be described as follows (Fig. 3.1B-D): *i*) three-layered covering membrane; *ii*) relatively electron transparent cortical layer; *iii*) thin granular zone projecting into the cortical layer at regular intervals; *iv*) triple layered zone, consisting of two fibre layers separated by a granular zone; *v*) more or less granular zone, containing several small lacunae; *vi*) fibrillar zone, subdivided into 7-12 lamellae, within which striation with intervals of *ca* 0.12 μm may be visible; and *vii*) an amorphous basal layer.

The anterior body region also contains an infracuticle, consisting of two parts: a thick zone containing numerous fine fibres, many thread-like epithelial projections lying in an obliquely longitudinal direction, and some lacunae; and a thin, amorphous basal layer to which the epidermis is attached (Roggen *et al.*, 1967; Aboul-Eid, 1969; Lippens *et al.*, 1974). Maggenti (1979) reviewed the nomenclature of the different layers and distinguished four strata (epi-, exo-, meso- and endocuticle) and pointed out that "among Enoplia the presence of all strata is remarkably constant. The mesocuticle varies by the number of fiber layers, their direction, and angular relationship. . . . The thickness of endocuticle is also variable".

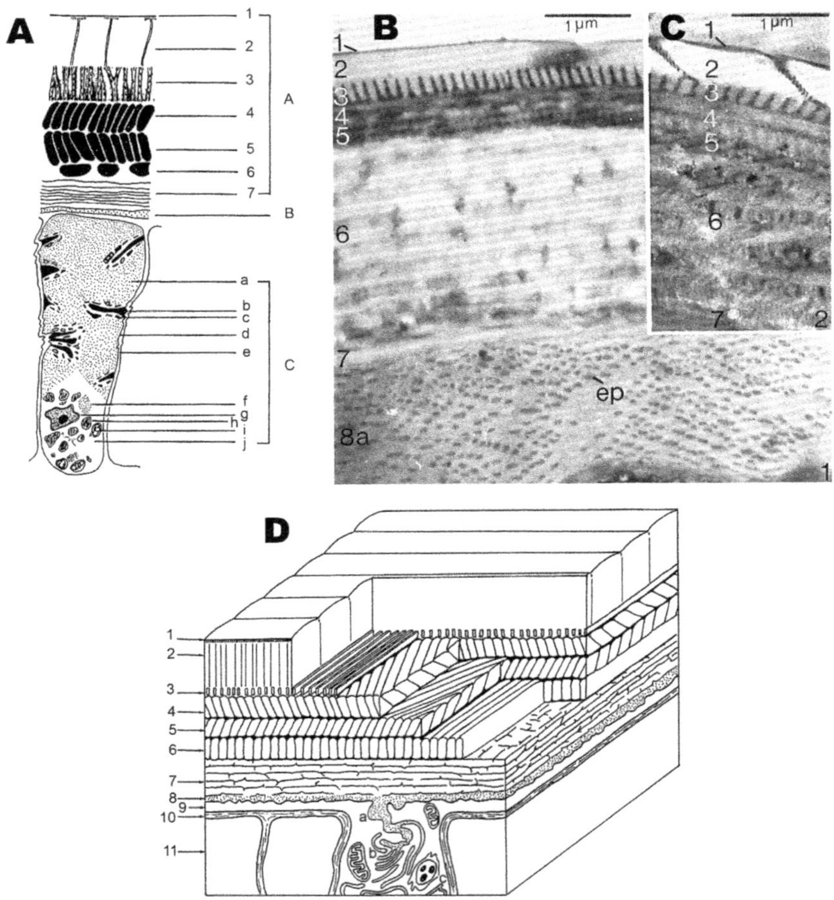

Fig. 3.1. *A: Schematic drawing of ultrastructure of body wall. Aa: Body cuticle; Ab: Epidermis; Ac: A body muscle cell. 1-7 = cuticle layers, a = myo-filaments, b = thickening, c = differently orientated thickening, d = sarcolemma, e-i = cell organelles, j = sarcoplasm (after Aboul-Eid (1969),* Nematologica *15, p. 459, Fig. 4); B, C: Oblique cross-section of cuticle showing its layering under TEM. 1-6 = cuticle layers (after Lippens* et al. *(1974),* Nematologica *20, p. 246, Fig. 1); D: Schematic drawing representing structure of body wall of* Xiphinema index*: 1-8 – the layers of the cuticle; 9 – epidermis; 10 – basement membrane; 11 – somatic muscle; a – invaginations of cell membrane in lateral epidermal chords; b – endoplasmic reticulum; c – cross section through nerves innervating lateral somatic papilla (after Roggen* et al. *(1967),* Nematologica *13, p. 2, Fig. 1).*

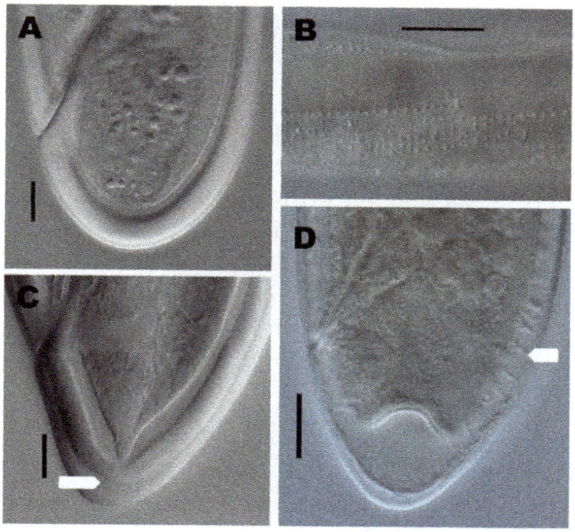

Fig. 3.2. *A: Dorylaimoid two-layered cuticle; B: Radial elements of leptonchid or tylencholaimoid cuticle in submedian view; C: Dorylaimoid, three-layered cuticle; arrow head pointing at 'empty' space between inner and intermediate layers; D: Leptonchid or tylencholaimid cuticle (arrow head pointing at radial refractive elements). (A:* Labronema neopacificum *from Vietnam; B:* Proleptonchus *sp. from Spain; C:* Aporcelaimellus obtusicaudatus *from Spain; D:* Chitwoodiellus parafuscus *from Spain.) (Scale bars: A, C, D = 10 μm; B = 20 μm.)*

Light microscopy shows, however, a significantly simpler structure, but also displaying some remarkable variations. In the most common scheme, the cuticle is made up of two (outer and inner) adjacent layers (Fig. 3.2A). It is the typical dorylaimoid cuticle pattern: a thin outer layer with constant thickness throughout the entire body, and an inner layer almost always thicker than the outer and becoming especially widened in the caudal region. The members of the important families Leptonchidae and Tylencholaimidae display a peculiar cuticle design, termed leptonchid or tylencholaimid, with their inner layer irregularly outlined and detached from the outer one, and having abundant small radial refractive elements in between (and apparently connecting) both cuticle layers (Fig. 3.2B, D). These radial elements appear in median view as small rodlets (Fig. 3.2D), but in submedian view they appear as rows of small dots (Fig. 3.2B).

A three-layered cuticle is characteristic of the cosmopolitan genus *Aporcelaimellus* (Fig. 3.2C), but it is also occasionally observed in other taxa. In this case, the cuticle consists of a thicker and more refringent inner layer, often bearing radial striation, and two (intermediate and outer) thinner layers. In addition, *Aporcelaimellus* species are distinguishable by the existence of an 'empty' or hyaline space between the inner and the intermediate layers at the tail end.

Sood & Kalra (1977) found that the cuticle of *Xiphinema insigne* is rich in sudanophilic lipids and that it also contains weakly acidic mucopolysaccharides.

Surface ornamentation

The external surface of the cuticle usually shows very fine transverse striation, often hardly discernible by means of LM, under which it appears nearly smooth, but sometimes well distinct (Fig. 3.3A). Striation is, however, easily visible under SEM and consists of delicate transverse (circular) incisures that usually present a regular (parallel) arrangement at the mid-body region (Fig. 3.3B), whereas they become more irregular (with frequent anastomoses) at the anterior (Fig. 3.3C) and posterior regions (note that cuticle striation in dorylaims is not at all comparable to that observed in other nematodes like tylenchs or rhabditids, in which 'true' annuli are easily perceptible). A few genera of the families Actinolaimidae and Dorylaimidae are characterised by having very prominent longitudinal ridges (Fig. 3.3D), running from the cervical region to the caudal region, whose number, especially distinct in body transverse sections (Fig. 3.3E), is of taxonomic interest. Exceptionally, the extremely rare genera *Encholaimus* and *Helmabia* bear deep transverse and longitudinal striation (incisures), the cuticle as a result appearing divided into blocks (Fig. 3.3F) throughout the entire body.

Cuticle irregularities of variable nature (wrinkles, folds, holes, flaps, *etc.*) may be present either close to the vulva or on the ventral side of the tail in a few taxa. Members of the genus *Rhyssocolpus* are characterised by having strongly developed cuticle irregularities in the vulva area, sometimes extending around the whole body perimeter (Fig. 3.3G-I). Some species of several genera (*Crassolabium*, *Dorylaimoides*, *Eudorylaimus*, *Longidorella*, *etc.*) bear especially coarse striation or some kind of irregularity on the ventral contour of the tail.

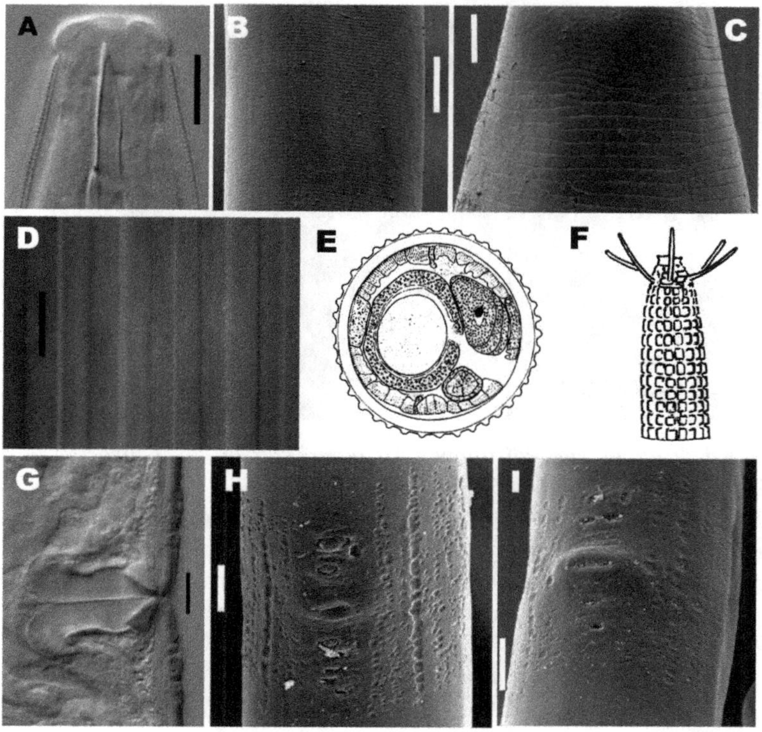

Fig. 3.3. *Cuticle surface. A: Cuticle bearing fine but perceptible transverse striation; B: Cuticle bearing fine but perceptible transverse striation at mid-body; C: Cuticle bearing fine but perceptible transverse striation in anterior body region (SEM); D: Cuticle bearing longitudinal ridges (LM); E: Cuticle bearing longitudinal ridges as seen in transverse section; F: Cuticle in blocks; G: Cuticle irregularities near vulval area as seen in LM lateral view; H: Cuticle irregularities near vulval area as seen in SEM ventral view; I: Cuticle irregularities near vulval area as seen in SEM ventral view. (A:* Aporcelaimellus salicinus *from Spain (LM). B, C:* Axonchoides smokyensis *from the USA (SEM). D:* Dorylaimus *sp. from Spain. E:* Dorylaimus stekhoveni *from the former Zaire (after Baqri & Coomans (1973),* Exploration du Parc National des Virunga, *p. 12, Fig. 3J). F:* Helmabia vera *(after Siddiqi (1971),* Zoologischer Anzeiger *186, p. 370, Fig. 1A). G-I:* Rhyssocolpus iuventutis *from Spain.) (Scale bars: A, B, D, G-I = 10 μm; C = 2 μm.)*

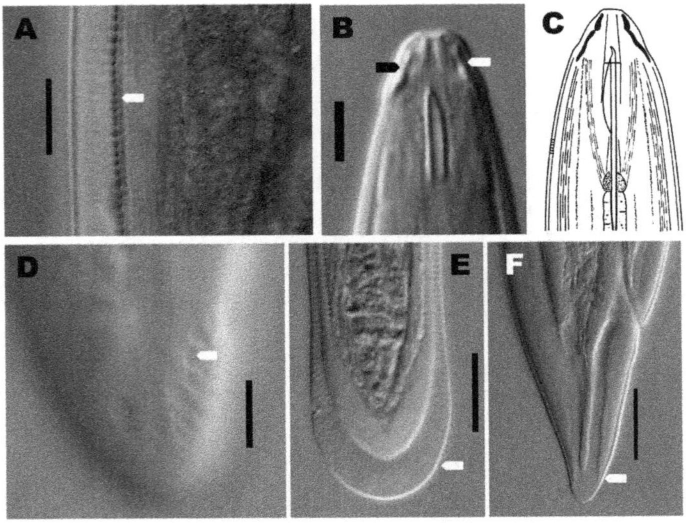

Fig. 3.4. *Cuticle specialisations. A: Inner cuticle layer bearing distinct striation; B, C: Labial and postlabial sclerotisation; D: Saccate bodies in cuticle of caudal region; E: Outer cuticle layer especially thickened in caudal region; F: Cuticle lacking inner layer in caudal region tip. (A:* Idiodorylaimus *sp. from Argentina (LM). B:* Belondira *sp. from Vietnam. C:* Margollus hispanicus *from Iberian Peninsula (after Peña-Santiago* et al.*, 1993,* Nematologica *39, p. 229, Fig. 4C). D:* Crassolabium circuliferum *from Romania. E:* Belondira *sp. from Vietnam. F:* Metaporcelaimus capitatus *from the USA.) (Scale bars: A, F = 20 μm; B, D, E = 10 μm.)*

Specialisations

Other differentiations of the typical dorylaimid cuticle may be seen in particular taxa. Thus, the existence of criss-cross lines has been reported in *Aporcelaimus* species, which are characterised by having an especially thick inner cuticle whose fibrous elements display a perpendicular orientation. Cuticle punctations have been described in members of *Makatinus* and *Aporcelaimellus*, although they certainly correspond to radial striation when observed in submedian view (*cf.* De Ley *et al.*, 1993). Exceptionally, the inner cuticle layer may have stronger transverse striation than the outer one in *Lordellonema* and its relatives, and also in *Idiodorylaimus* (Fig. 3.4A).

More local (regional) specialisations also exist. The presence of labial and/or postlabial sclerotisations (especially refractive cuticle areas at the lip region and the adjoining body) form a cephalic framework, an interesting diagnostic feature for members of the subfamily Thornenematinae, but also known to occur in other genera, for example, *Belondira* and *Margollus* (Fig. 3.4B, C). The caudal region may present other kinds of differentiations. It is not unusual to find the presence of bubble-like or saccate bodies inside the cuticle (Fig. 3.4D), mostly on the ventral side, as has been repeatedly described in representatives of the genera *Crassolabium*, *Enchodelus*, *Eudorylaimus*, *Pungentus* and others. The outer layer may become especially developed on the tail, forming a large hyaline area in some species of the genus *Belondira* and other belondirid taxa (Fig. 3.3E). Lastly, the inner layer may have a discontinuity or gap at the tail end, as often reported in members of *Metaporcelaimus* (Fig. 3.4F).

Body pores

The body cuticle is penetrated by thin canals (Fig. 3.5A) leading to small orifices (pores) on the body surface. These canals originate from the four epidermal chords and their function has not yet been elucidated, although they may represent small ducts of the epidermal glands or even have some sensory role (Jairajpuri & Ahmad, 1992). Their nature, arrangement and number vary among dorylaimid taxa. They are often small, simple and dot-like (Fig. 3.4A), or appear as low depressions of the body surface; more rarely they become larger and are encircled by a ring-like elevation of cuticle (Fig. 3.5E). Body pores are usually distributed in one row, occasionally two rows (Fig. 3.5B), throughout the lateral chords of the entire body, often starting posterior to the lip region and ending anterior to the caudal region. Many species also possess ventral and dorsal body pores, these being fewer in number than the lateral pores and restricted to the anterior body region, hence the term cervical pores (Fig. 3.5A, C). These dorsal and ventral pores are easily seen due to their fine canals crossing the cuticle when the observer looks at the nematode in lateral view (Fig. 3.5C). A row of ventral pores along the entire body has also been noted in some species, sometimes with their corresponding canals especially conspicuous (Fig. 3.5D). The number of lateral and/or ventral pores is highly variable, ranging from a few tens to several hundred.

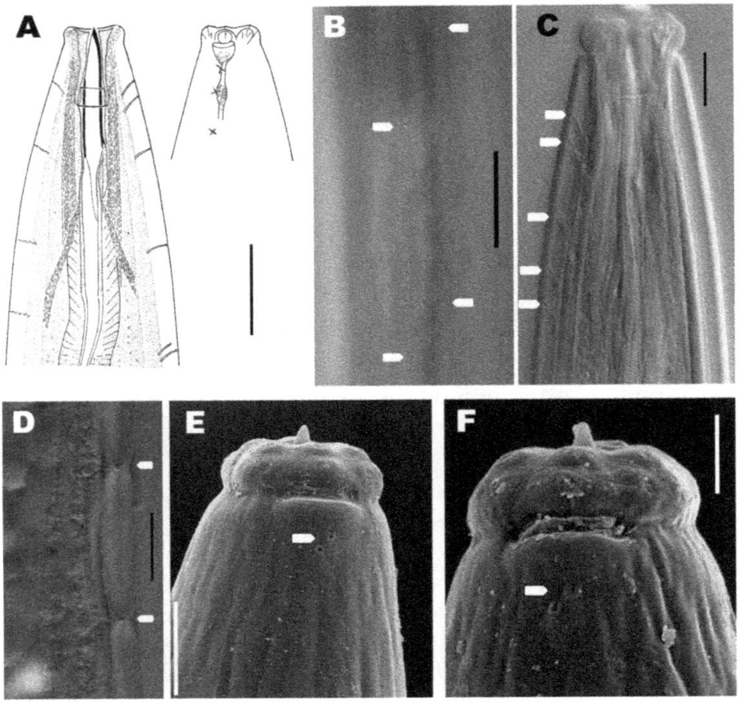

Fig. 3.5. *Body pores. A: Four dorsal, six ventral and three lateral body pores in anterior body region; B: Four lateral pores, arranged in two rows; C: Location of five ventral body pores and their corresponding ducts in anterior region; D: Two coarse ventral body pores and their corresponding ducts in mid-body; E: Two lateral body pores posterior to amphidial aperture. (A:* Dorylaimus *sp. from the former Zaire (after Baqri & Coomans (1973),* Exploration du Parc National des Virunga, *p. 16, Fig. 5A). B:* Rhyssocolpus vinciguerrae *from Iran. C:* Labronema carusoi *from Spain. D:* Aporcelaimellus porosus *from Guinea. E, F:* Nevadanema nevadense *from Spain.) (Scale bars: A = 50 μm; B, D = 20 μm; C, E = 10 μm; F = 5 μm.)*

Epidermis and lateral chords

According to Lippens *et al.* (1974), in *Aporcelaimellus* species and as derived from TEM observations, the epidermis is a sheath of vacuolated tissue lying beneath the cuticle (infracuticle at the anterior region) and attached to this by desmosome-like structures. Upon fixation, the epidermis often becomes somewhat detached from the body cuticle, showing an artificial, 'empty' space (*lacuna*), which is not a cuticle layer

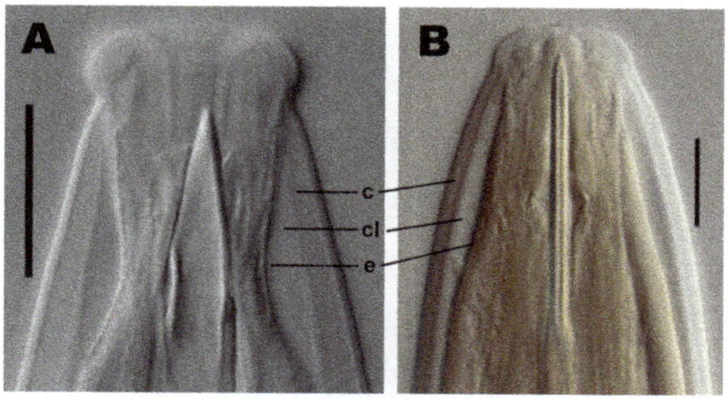

Fig. 3.6. *Cervical lacunae. A, B: Cervical lacuna (lc) between cuticle (c) and epidermis (e). (A:* Aporcelaimellus californicus *from USA. B:* Enchodelus longispiculus *from Spain.) (Scale bars = 10 μm.)*

but an artefact due to tissue contraction (*cf.* Lippens & Grootaert, 1974; Lippens *et al.*, 1974; De Ley *et al.*, 1993). Such a *lacuna* is, however, readily observable in the anterior region of the body, reaching particular development in several dorylaimid groups (Fig. 3.6A, B), in which they might represent structural elements, even perceptible in live specimens – Andrássy (2002) actually regarded their presence as a diagnostic feature for *Aporcelaimellus* species.

Under both TEM and LM, a transverse section of the body reveals that the epidermis bulges into the pseudocoel in four (one ventral, one dorsal and two lateral) epidermal chords (Fig. 3.7A). These chords anteriorly consist of some epithelial tissue as well as important nervous and glial elements. Epithelial cell nuclei in the chords are situated underneath the infracuticle, but their location and arrangement seem variable. The two (right and left) lateral chords are more developed than the dorsal and ventral chords and can be easily seen beneath the cuticle forming a longitudinal band throughout most of the body length (Fig. 3.7B). Their width is a rather constant feature among members of the same species but variable when representatives of different genera or families are compared, ranging from one-tenth to one-third of the mid-body diameter. They often present a granular appearance, but lack any particular differentiation. Nevertheless, lateral chords occasionally contain gland bodies arranged in a longitudinal series (Fig. 3.7C), each body apparently leading to a lateral pore, as observed in species of the subfamily Discolaiminae and more sporadically in particular species of several genera.

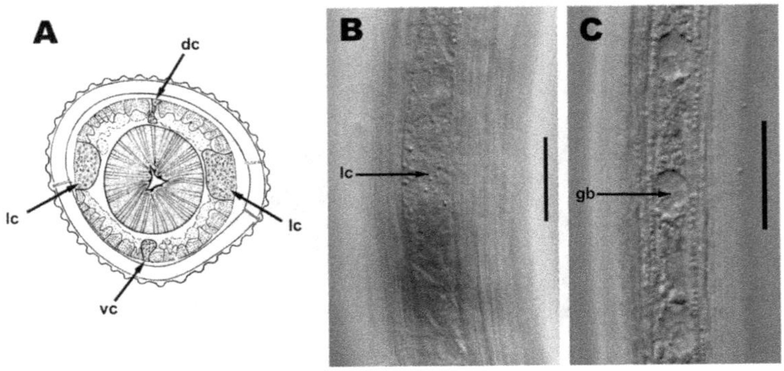

Fig. 3.7. *Epidermis and chords. A: Transverse section at level of pharynx showing dorsal (dc), ventral (vc) and lateral (lc) chords; B: Lateral chord (lc) in lateral view; C: Lateral chord in lateral view, showing gland bodies (gb). (A:* Dorylaimus stekhoveni *from the former Zaire (after Baqri & Coomans (1973),* Exploration du Parc National des Virunga, *p. 12, Fig. 31). B:* Rhyssocolpus iuventutis *from Spain. C:* Aporcella vitrinus *from Spain.) (Scale bars = 20 μm.)*

Somatic musculature

The somatic musculature is separated from the epidermis by an area of connective tissue, the basal lamina (Lippens, 1975). It is either polymyarian with a coelomyarian type of muscle cells or meromyarian with a platymyarian type of muscle cells (Jairajpuri & Ahmad, 1992). The epidermal chords divide the muscle layer into four sectors. TEM observations of *Longidorus macrosoma*, however, show that the four primary muscle fields (sectors) are further divided by subdorsal and subventral chords into eight groups of cells (Aboul-Eid, 1969). Each of the four main sectors is made of 10-12 muscles at the mid-body region, but this number decreases towards the anterior and posterior ends (Das, 1963).

As revealed by TEM in *Aporcelaimellus* species, the musculature is coelomyarian, composed of a single layer of cells longitudinally arrayed (Lippens *et al.*, 1974; Lippens, 1975). The fibrillar part contains two types of filaments as well as T-tubules accompanied by smooth endoplasmic reticulum. The sarcoplasmic part is rich in mitochondria and ribosomes and, especially, in glycogen; some cisternae of rough endoplasmic reticulum are also present.

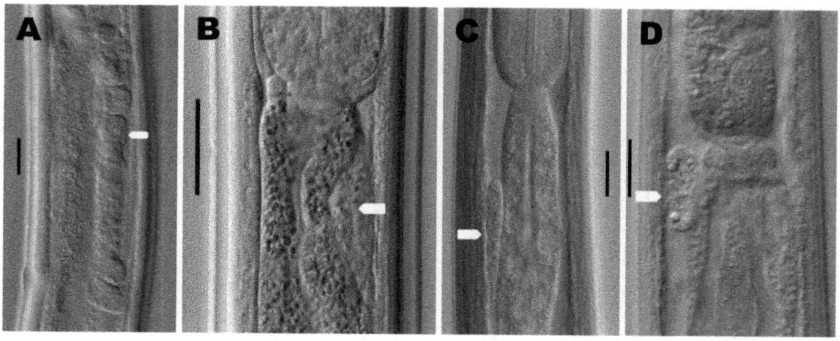

Fig. 3.8. *Dorsal cell chord (A) or mass (B-D) at anterior end of intestine. A:* Crassolabium vietnamense *from Vietnam; B:* C. rhopalocercum *from Iran; C:* Enchodelus carpaticus *from Romania; D:* Oriverutus ecae *from Costa Rica. (Scale bars: A-C = 20 μm; D = 10 μm.)*

In addition to the muscle layer forming part of the body wall, there is other specialised musculature situated in the anterior and posterior regions as well as at the vaginal region of the female. These special muscles will be described and discussed in the corresponding sections.

Pseudocoel and its components

In *Aporcelaimellus* species, the body cavity (pseudocoel) is packed with granular and fibrous material with some cells in it that histologically correspond to a loose-fibrillar connective tissue (Lippens *et al.*, 1974).

The existence of pseudocoelomocytes has not been reported in dorylaims. Nevertheless, the presence of a well-developed dorsal cell mass (up to three times as long as wide; Fig. 3.8B-D) or chord (more than three times as long as wide; Fig. 3.8A), apparently floating within the body cavity, is characteristic of a few genera, for example *Paravulvus* and *Aporcelinus*, and more occasionally observed in species of many others. This cell mass/chord usually measures a few tens of microns long and is located at or slightly posterior to cardia level at the anterior end of the intestine, whose wall often presents a concavity within which the mass/chord is partially lodged.

References

ABOUL-EID, H.Z. (1969). Electron microscope studies on the body wall and feeding apparatus of *Longidorus macrosoma*. *Nematologica* 15, 451-463. DOI: 10.1163/187529269X00786

ANDRÁSSY, I. (2002). Free-living nematodes from the Fertö-Hanság National Park, Hungary. In: Mahunka, S. (Ed.). *The fauna of the Fertö-Hanság National Park*. Budapest, Hungary, Hungarian Natural History Museum, pp. 21-97.

BAQRI, Q.H. & COOMANS, A. (1973). A taxonomic revision of the nematode species described by S. Stekhoven & Teunissen (1938) and S. Stekhoven (1944) from National Virunga Park, Zaïre Republic. I. Dorylaimidae, Aporcelaimidae and Longidoridae. *Exploration du Parc National des Virunga* 1, 3-57.

DAS, V.M. (1963). Morphological studies of the male *Nygolaimus aquaticus* Thorne, 1930. *Canadian Journal of Zoology* 41, 725-732.

DE LEY, P., LOOF, P.A.A. & COOMANS, A. (1993). Terrestrial nematodes from the Galápagos Archipelago II: Redescription of *Aporcelaimellus obtusicaudatus* (Bastian, 1865) Altherr, 1968, with review of similar species and a nomenclature for the vagina in Dorylaimida (Nematoda). *Bulletin de l'Institut Royal des Sciences Naturelles de Belgique* 63, 13-34.

HEYNS, J. & JACOBS, P.J.F. (1989). Comparative ultrastructure of the cuticle in species of *Longidorus, Longidoroides, Paralongidorus* and *Xiphinema* (Nematoda: Longidoridae). *Phytophylactica* 21, 265-270.

JAIRAJPURI, M.S. & AHMAD, W. (1992). *Dorylaimida. Free-living, predaceous and plant-parasitic nematodes*. Leiden, The Netherlands, Brill.

LIPPENS, P.L. (1975). Ultrastructure of Dorylaimidae. In: Lamberti, F., Taylor, C.E. & Seinhorst, J.W. (Eds). *Nematode vectors of plant viruses*. London and New York Plenum Press and NATO Scientific Affairs Division, pp. 159-178.

LIPPENS, P.L. & GROOTAERT, P. (1974). A routine method for mounting nematodes in resin with high refractive index. *Nematologica* 19(1973), 562-563. DOI: 10.1163/187529273X00592

LIPPENS, P.L., COOMANS, A., DE GRISSE, A.T. & LAGASSE, A. (1974). Ultrastructure of the anterior body region in *Aporcelaimellus obtusicaudatus* and *A. obscurus*. *Nematologica* 20, 242-256. DOI: 10.1163/187529274X00212

MAGGENTI, A. (1979). The role of cuticular strata nomenclature in the systematics of Nemata. *Journal of Nematology* 11, 94-98.

PEÑA-SANTIAGO, R., PERALTA, M. & SIDDIQI, M.R. (1993). Taxonomy of some new and known species of the genus *Tylencholaimellus* with a proposal for *Margollus* gen. n. (Nematoda: Dorylaimida). *Nematologica* 39, 218-233. DOI: 10.1163/187529293X00178

ROGGEN, D.R., RASKI, D.J. & JONES, N.O. (1967). Further electron microscopic observations of *Xiphinema index*. *Nematologica* 13, 1-16. DOI: 10.1163/187529267X00850

SIDDIQI, M.R. (1971). *Helmabia*, a new genus of dorylaimid nematodes with cephalic setae. *Zoologischer Anzeiger* 186, 368-372.

SOOD, M.L. & KALRA, S. (1977). Histochemical studies on the body wall of nematodes: *Haemonchus contortus* (Rud., 1803) and *Xiphinema insigne* Loos, 1949. *Zeitschrift für Parasitenkunde* 51, 265-273.

Nematology Monographs & Perspectives, 2021, Vol. 13, 27-39

Chapter 4

Lip region and amphids

The anterior body end of nematodes is termed the lip (labial) region (or head). In dorylaims, its posteriormost limit is not easy to establish, although it is often assumed that the amphidial aperture (see below) is situated just posterior. The mouth (*i.e.*, the oral aperture) is located in the centre of the anteriormost surface. The lip region shows the typical hexaradial nematode symmetry and displays a characteristic pattern of 6 + 6 + 4 sensilla. The lip region morphology displays a number of important and taxonomically relevant features.

Lip region shape

Several morphological features are involved in the characterisation of the lip region in dorylaims: profile, anterior margin, tapering, and degree of separation from the adjoining body. These will be dealt with separately.

PROFILE (CONTOUR)

The profile or contour ranges from totally smooth to conspicuously angular. The lip region may show a smooth contour (Fig. 4.1A) as the lips become amalgamated and the sensilla hardly protrude above their surface. Much more often, however, it becomes variably angular since the lips appear separate and/or the sensilla noticeably protrude above their surface. Nonetheless, the degree of separation in the lips and protrusion of the sensilla is extremely variable among species, with many intermediate possibilities, from weakly (Fig. 4.1B) to distinctly (Fig. 4.1C) angular forms.

ANTERIOR MARGIN

The anterior margin may be rounded, more or less convex, and visibly higher in the vicinity of the inner (perioral) area than the outer

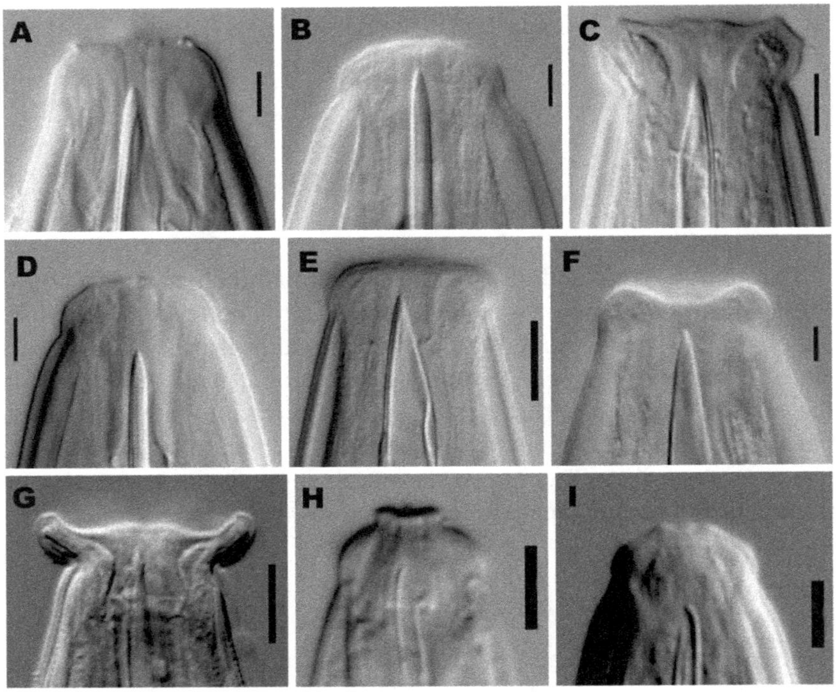

Fig. 4.1. *Morphology of lip region as seen under LM (part I). A-C: Profile; D-F: Anterior margin; G-I: Special shapes. A: Smooth profile; B: Weakly angular profile; C: Angular profile; D: Anterior margin rounded; E: Anterior margin truncate or flat; F: Anterior margin depressed; G: Sucker-like; H, I: Cap-like. (A:* Labronema neopacificum *from Vietnam. B, D:* Enchodelus saxifragae *from Romania. C:* L. carusoi *from Romania. E:* Aporcella simplex *from Spain. F:* Dorylaimus *sp. from Spain. G:* Discolaimus major *from Spain. H:* Tyleptus *sp. from Vietnam. I:* Tylencholaimellus auringiensis *from Spain.) (Scale bars: A-D, F, H, I = 5 μm; E, G = 10 μm.)*

(peripheral) part (Fig. 4.1D). More usually, however, the anterior margin appears truncate or flat, being equally high at its inner and outer parts (Fig. 4.1E). More rarely, the lip region appears depressed, as its inner region is perceptibly lower than its outer part (Fig. 4.1F), and may even become sucker-like in extreme cases (Fig. 4.1G). It is not unusual for the inner area of the lip region to become distinctly more protruding and higher than the outer part, this condition being referred to as cap-like (Fig. 4.1H, I).

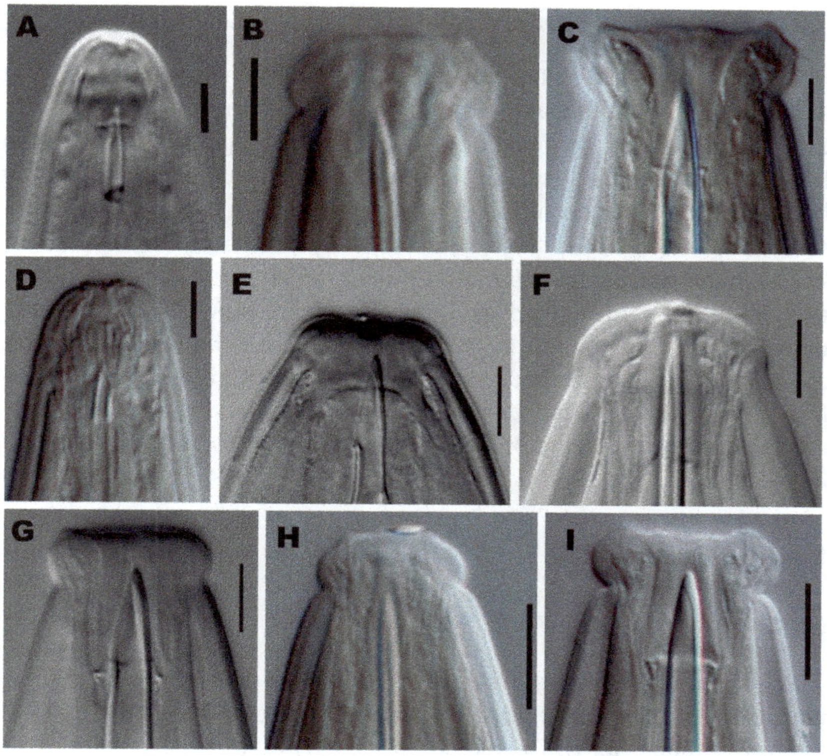

Fig. 4.2. *Morphology of lip region as seen under LM (part II). A-C: Lip region tapering. D-I: Lip region separation from the adjoining body. A: Gradually tapering; B: Non-tapering; C: Slightly expanded; D, E: Continuous; F: Offset by weak depression; G: Offset by marked depression; H: Offset by weak constriction; I: Offset by deep constriction. (A:* Belondira *sp. from Vietnam. B:* Longidorella deliblatica *from Spain. C:* Labronema carusoi *from Romania. D:* Adenolaimus orthus *from Iran. E:* Makatinus minor *from Sri Lanka. F:* Enchodelus groenlandicus *from Iran. G:* Labronema *sp. from Vietnam. H:* Heterodorus geraldi *from Romania. I:* Labronema angeloi *from Spain.) (Scale bars: A-D = 5 μm; E-I = 10 μm.)*

TAPERING

Lip region width (diameter) is often equal to that of the adjoining body (Fig. 4.2B). In some cases, however, it gradually narrows, even within the short distance from its base to the anterior margin (Fig. 4.2A). More sporadically, it can be expanded, that is, slightly (Fig. 4.2C) to distinctly (Fig. 4.1G) wider than the adjoining body.

DIFFERENTIATION (SEPARATION FROM THE ADJOINING BODY)

It is not unusual for the lip region to be perfectly continuous (Fig. 4.2A, D, E) with the adjoining body, its posteriormost limit being established only by the location of the amphidial apertures. Very often, however, it becomes offset, *i.e.*, marked or separated from the adjoining body by the existence of either a depression or a constriction. If the separation between lip region and its adjacent body is marked off by a slight narrowing (Fig. 4.2F) or a weak incisure (Fig. 4.2G), this condition is described as 'offset by depression'. When a strong narrowing exists, marked by a deep incisure separating both portions of the body, it is described as 'offset by a constriction' (Fig. 4.2H, I). Many intermediate forms can occur and interpretation is influenced by the subjective assessment of the observer and/or the preservation state of the specimens.

Lips and their papillae

GENERAL PATTERN

As mentioned above, the dorylaims follow the general nematode body plan, their lip region exhibiting hexaradial symmetry (De Coninck, 1942, 1965; Coomans, 1978; Decraemer *et al.*, 2014). It consists of six (two lateral, two subdorsal and two subventral) lips, each bearing two (one inner and one outer) labial papillae. In addition, the subdorsal and subventral lips each bear a single cephalic papilla. Scanning electron microscopy (SEM) allows the observation of many details of the lips and their papillae (Fig. 4.3).

LIPS

Several features influence the morphology of the lips. Their degree of fusion/separation is important and useful for taxonomical purposes, but not always easy to appreciate under LM. Lips may become totally fused or amalgamated (Fig. 4.4A), and are only discernible by the presence of their papillae; many species, however, show partially amalgamated/separate lips (Fig. 4.4B) and, much less frequently, the lips are totally separate, even to the extent of forming independent lobes (Fig. 4.4C). It is not unusual that the lip region exhibits a perceptible biradial symmetry. This occurs when the lateral lips become visibly

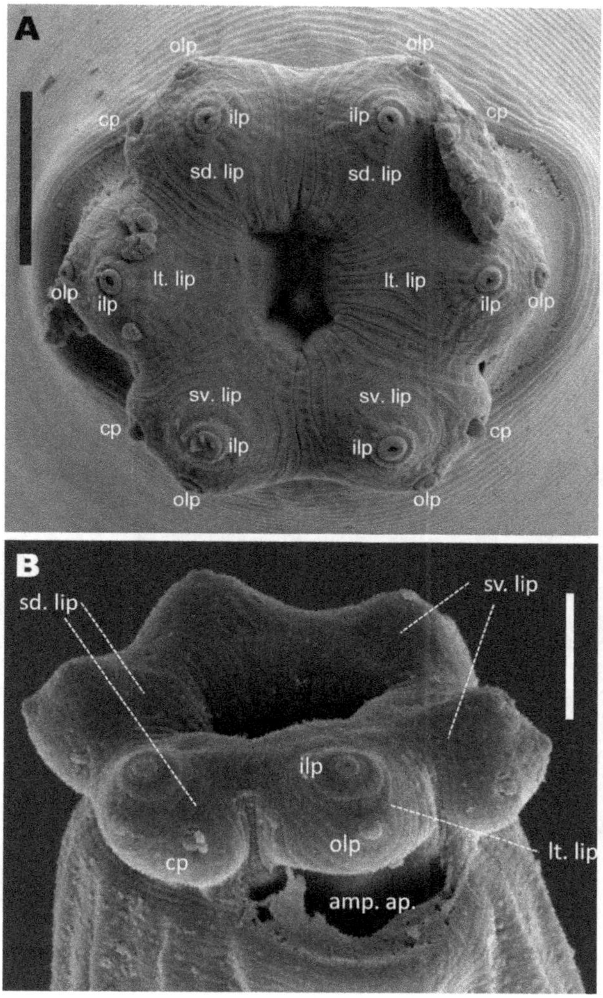

Fig. 4.3. *Lip region of* Aporcelaimellus communis *observed with SEM. A: Frontal view; B: Lateral view. Abbreviations: amp. ap. = amphid aperture; cp = cephalic papilla; ilp = inner labial papilla; lt. lip = lateral lip; olp = outer labial papilla; sd. lip = subdorsal lips; sv. lip = subventral lips. (Scale bars = 5 μm.)*

larger than the others (Fig. 4.4F) and/or when the lateral lips appear more marked from their corresponding subventral and subdorsal neighbours, the subventral and subdorsal lips remaining less separate between them (Fig. 4.4D, E).

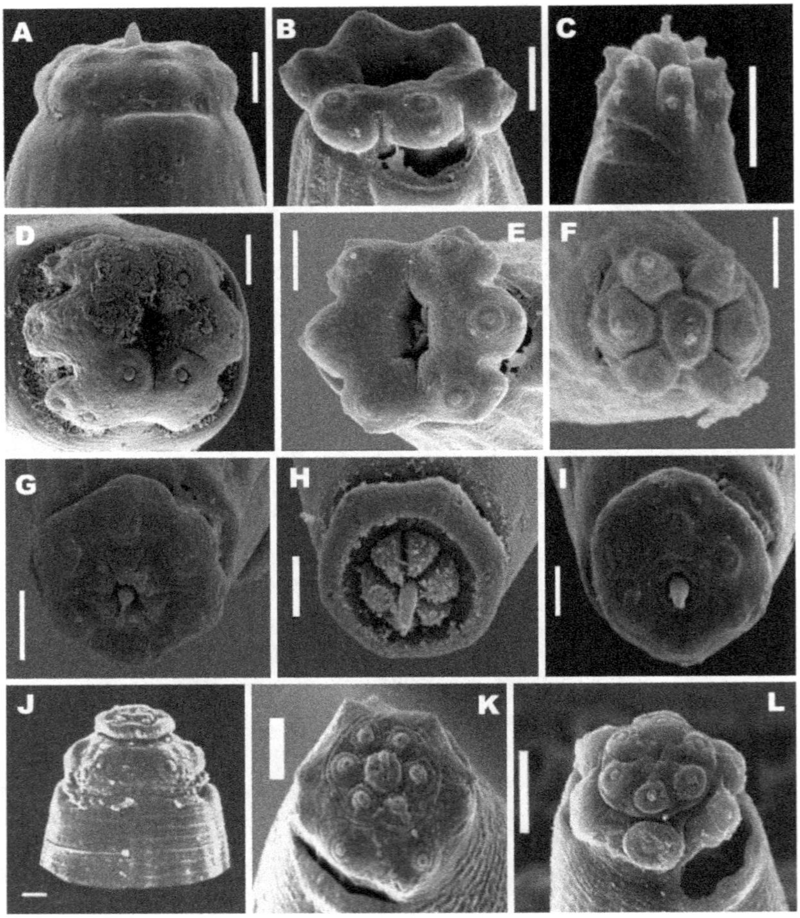

Fig. 4.4. *Lips and their papillae (SEM). A: Amalgamated; B: Partially separated; C: Separated; D-F: Lips arranged according to a biradial symmetry pattern; G: Low inner (perioral) liplets; H: Distinct and protruding inner (perioral) liplets; I: Ring-like perioral area; J: Disc-like perioral area; K: Inner labial papillae close to oral aperture; L: Offset perioral area (including inner labial papillae). (A:* Nevadanema nevadense *from Spain. B, E:* Aporcelaimellus communis *from Spain. C:* Oriverutus assacatus *from Costa Rica. D:* Axonchoides smokyensis *from the USA. F:* O. sundarus *from India. G:* Labronema angeloi *from Spain. H:* Labronemella labiata *from Iran. I:* N. nevadense *from Spain. J:* Tylencholaimus constrictus *from Spain. K:* Rhyssocolpus iuventutis *from Spain. L:* Chitwoodiellus parafuscus *from Spain.) (Scale bars: A-C, E, G-I = 5 μm; D = 2 μm; F, K, L = 3 μm; J = 10 μm.)*

The inner (perioral) region of the lips, *i.e.*, the area surrounding the mouth aperture, may not display any peculiar feature apart from the possible existence of radial striation starting from the oral aperture and ending at the lip region margin (Fig. 4.4E). In some cases, however, this inner region may be especially differentiated in several ways. The presence of six perioral liplets is characteristic of some genera. In *Labronema* species, for example, the liplets are low but offset and readily perceptible (Fig. 4.4G), whereas in *Labronemella* representatives they are very conspicuous and distinctly offset (Fig. 4.4H). A perioral, ring-like structure, delimited by a circular striation or incisure that is concentric with the oral aperture, is also reported in some species (Fig. 4.4I). Much less usually, the perioral region is perceptibly separate from the remaining lip region, thus forming a lobe, as happens in some *Oriverutus* species (Fig. 4.4C, F). Other special differentiations occur in many species of the families Belondiridae, Tylencholaimidae, Leptonchidae and Tylencholaimellidae, with a cap-like lip region whose inner region is distinctly protruding and forms an offset, disc-like, perioral structure (Fig. 4.4J-L).

PAPILLAE

Dorylaims exhibit the usual nematode pattern of 6 + 6 + 4 sensilla. They consist of a first circle of six inner labial sensilla surrounding (but at a variable distance from) the oral aperture, and a second circle of six outer labial and four cephalic sensilla located near the lip region margin; occasionally, the four cephalic sensilla may form a third circle situated a very short distance posterior to the level of the outer labial papillae (Fig. 4.5B). Labial sensilla occupy a radial position whereas cephalic sensilla are situated in a submedian position in the centre of each body quadrant (Fig. 4.3A). The sensilla are invariably papilliform or button-like. Very exceptionally, however, the representatives of some extremely rare genera, for example *Helmabia* or *Cephalodorylaimus*, bear long, cornuate, or setiform sensilla (setae) (Fig. 3.3F). Labial papillae each have a central pore. Cephalic papillae often bear a very short transverse slit, more seldom appearing as short, bevelled, tube-like structures. Papillae may appear especially conspicuous when the lip region lacks any striation, being then located either in depressions of the lip region surface (Fig. 4.4D) or projecting from this surface (Fig. 4.4C). Nevertheless, lip region surface is commonly sculptured by radial and/or circular striation and the papillae are more or less integrated into this scheme. Thus, each papilla, in particular the labial ones, are often surrounded by one or a few concentric incisures (Fig. 4.4E, G, I).

ORAL APERTURE (MOUTH)

In frontal view, and when observed with SEM, the oral aperture displays interesting variation that is often correlated with the morphology of the perioral area of lips and/or the symmetry of the entire lip region. Thus, it may be a comparatively small, circular opening (Fig. 4.4I), a short dorsoventral slit (Fig. 4.4F) or a star-like (when deep radial incisures are present at the perioral area; Fig. 4.4D, G) orifice. Members of the family Aporcelaimidae, however, exhibit a large, dorsoventral, slit-like aperture with a nearly hexagonal contour (Fig. 4.4E). Very exceptionally, the existence of abundant cilia-like projections covering the oral aperture has been observed in some *Sectonema* species (Heyns, 1965; Álvarez-Ortega & Peña-Santiago, 2019), but the precise nature and function of these structures have not yet been elucidated.

ORAL FIELD

The oral field is the surface of the lip region extending from the oral aperture to the circumference delimited by the inner labial papillae. As observed in frontal view, the oral field may fit two basic patterns. On the one hand, the inner labial papillae may migrate to the margins of the lip region, close to their respective outer labial papillae. In this case, the oral field is wide, with appreciable space between the oral aperture and inner labial papillae, and usually appears ornamented with radial striation (Fig. 4.4E, G, H). On the other hand, the inner labial papillae may be situated near the oral aperture and comparatively far from their respective outer papillae (Fig 4.4D, K, L). In this case, the oral field is comparatively reduced and may show a perceptible ring-like incisure marking a circular area containing the oral aperture and the inner labial papillae (Fig. 4.4K). Eventually, when sufficiently developed, this circular area may appear as a marked disc (see above; Fig. 4.4L).

Amphids

Basic structure

These sense organs are characteristic components of the nematode body plan. Coomans (1979) provided a very detailed study of both the structure and the function of amphids in nematodes, which will be treated in more detail in Chapter 7. Each amphid consists of three basic elements: fovea, amphidial canal and fusus (Fig. 4.5A). In dorylaims, the fovea (also termed the amphidial pocket) is an invagination of the body cuticle that forms a small cavity or pocket connected to the exterior through an aperture. It is the most visible part of the amphid when specimens are observed with LM as the amphidial canal and fusus are less conspicuous.

Position

The position of the amphids marks the posteriormost limit of the lip region. They are regarded as postlabial (Fig. 4.5B) with their aperture situated just posterior to the lip region and at the level of the labial depression/constriction when this exists. To be more precise, the amphidial aperture is situated a short (but appreciable) distance posterior to the level of the outer labial papillae. Very exceptionally, however, the amphidial aperture migrates forwards and occupies a more anterior (labial) position, as in members of the genera *Ecumenicus* (Fig. 4.5C) and *Kochinema*, the lip region then displaying a biradial symmetry (Peña-Santiago & Abolafia, 2007), a very peculiar apomorphic condition within the Dorylaimida[1].

Aperture

The aperture often appears as a transverse, straight and narrow slit (Fig. 4.5B). Sometimes, but more frequently than supposed, the aperture

[1] From a phylogenetic (cladistic) perspective, the postlabial position of the amphids has been interpreted as a primitive (plesiomorphic) condition (Adamson, 1987), known to occur in members of the class Enoplea, whereas its anterior (forward) migration is a derived (apomorphic) state that characterises the class Rhabditea.

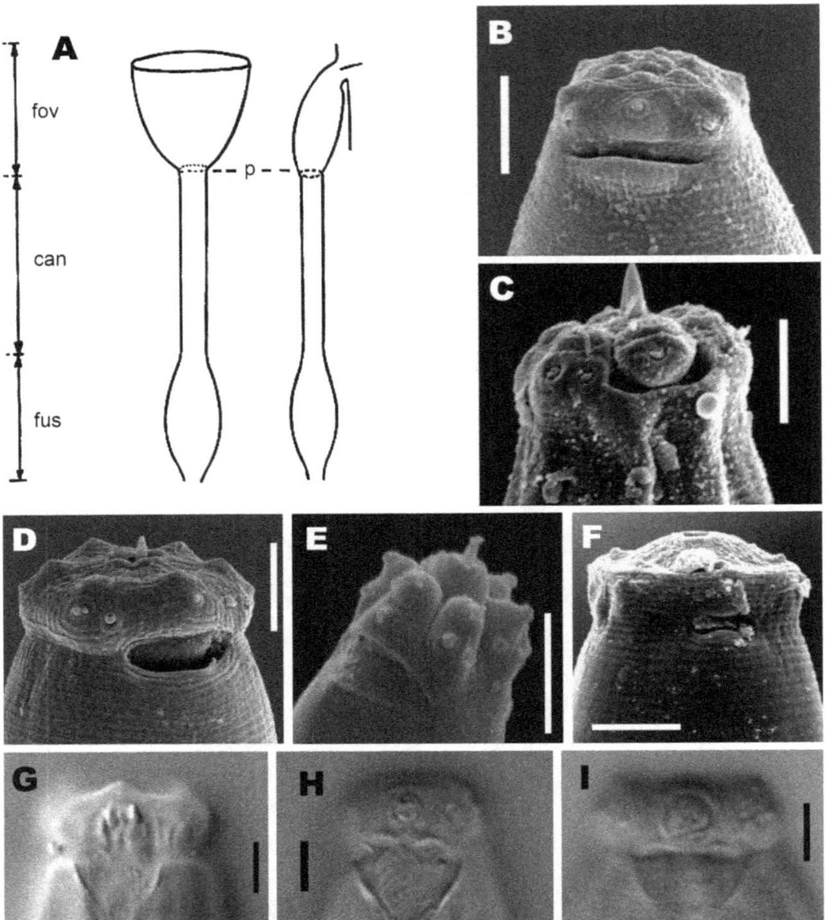

Fig. 4.5. *Amphids. A: General structure of the amphid in dorylaims (diagrammatic: can = canal; fov = fovea; fus = fusus; p = canal pore or duct pore). (After Coomans (1979),* Revue de Nématologie *2, p. 272, Fig. 4B.); B: Amphid (fovea) aperture in postlabial position; C: Amphid (fovea) aperture in anterior (labial) position; D-F: Variation of amphid (fovea) aperture as observed with SEM; G-I: Examples of fovea shape as seen with LM. (B:* Rhyssocolpus iuventutis *from Spain. C:* Ecumenicus monohystera *from Spain. D:* Crassolabium tenuistylum *from Romania. E:* Oriverutus asaccatus *from Costa Rica. F:* Pachydorylaimus notabenus *from Costa Rica. G:* Labronema angeloi *from Spain. H:* Sectonema *sp. from Vietnam. I:* Aporcelaimellus rotundicaudatus *from Sri Lanka.) (Scale bars: B, D-I = 5 μm; C = 3 μm.)*

is elliptical (Fig. 4.5D) or fusiform (spindle-like) (Fig. 4.5E). Exceptionally, it becomes very small and pore-like, as described in *Tantunema* species and some representatives of the family Longidoridae. Its (anterior and posterior) margins may be straight, but they usually appear more or less undulate, sometimes partially closing the opening at the middle (Figs 4.4L; 4.5F). Its size is highly variable, occupying in general one-third to one-half of the corresponding body diameter, seldom less and often more than this range, even extending for almost the entire lip region diameter.

FOVEA

Under the light microscope, and in lateral view, the fovea is readily visible across or beneath the cuticle. It usually appears as a cup-shaped cavity[2], almost as long as wide and with curved margins converging at the base (Fig. 4.5G-I). Nevertheless, it displays some remarkable variations. In longidorid species (but also a valid statement for dorylaims in general), Decraemer & Coomans (2007) distinguished four common (plus some other less common) types of amphids depending on the shape of their fovea (Fig. 4.6): *i*) stirrup-shaped, wider than long, aperture a transverse slit occupying more than one-half of the lip region diameter and curved walls tapering posteriorly; *ii*) open goblet-shaped, equally or longer than wide, aperture a transverse slit occupying up to one-half of the lip region diameter but never minute or pore-like, and curved walls tapering posteriorly; *iii*) funnel-shaped, straight walls tapering posteriorly, with variably sized aperture but never minute, and variable length and width; and *iv*) pouch-like, wider than long or longer than wide, with minute, pore-like opening, the posterior end rounded or bilobed, the latter type bearing symmetrical or asymmetrical lobes.

[2] The existence of a 'duplex' amphid, with a chamber-like structure posterior to the fovea, has been recorded in some species, for example, in representatives of the genus *Aporcelaimellus*. This tentative variation in amphid structure was discussed in some detail by De Ley *et al.* (1993), who concluded that further (ultrastructural) studies would confirm its status, but that it was very probably due to fixation effects on the somatic muscles.

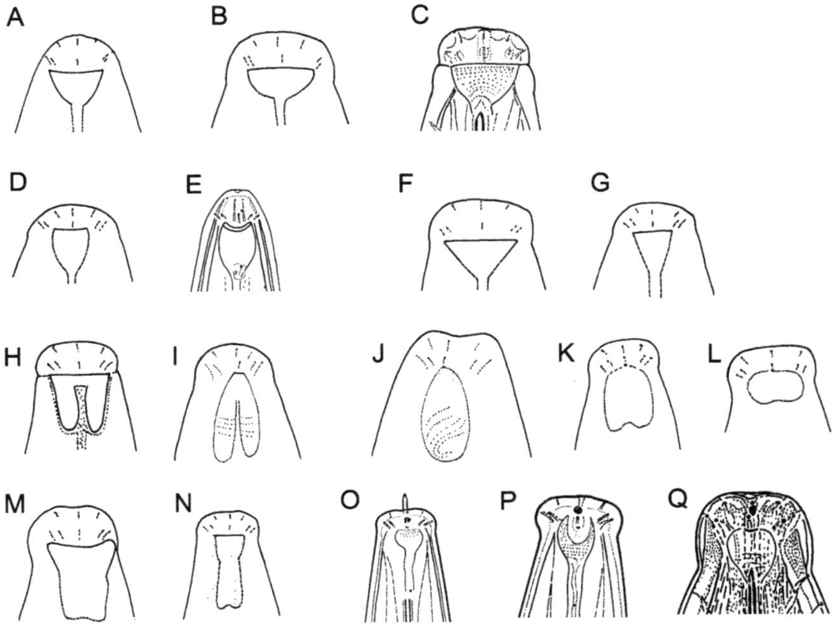

Fig. 4.6. *Different types of amphidial fovea (after Decraemer & Coomans (2007),* Nematology *9, p. 647, Fig. 1). A-C: Stirrup-shaped; D, E: Open goblet-shaped; F, G: Funnel-shaped; H-L: Pouch-like; M, N: Elongate stirrup-shaped; O, P: Closed goblet-shaped; Q: Barrel- or heart-shaped.*

References

ADAMSON, M.L. (1987). Phylogenetic analysis of the higher classification of the Nematoda. *Canadian Journal of Zoology* 65, 1478-1482.

ÁLVAREZ-ORTEGA, S. & PEÑA-SANTIAGO, R. (2019). Morphology, phylogeny and taxonomy of the genus *Sectonema* (Nematoda, Aporcelaimidae). *Zoologica Scripta* 48, 535-544. DOI: 10.1111/zsc.12357

COOMANS, A. (1978). A proposal for a more precise terminology of the body regions in the nematodes. *Annales de la Société Royale Zoologique de Belgique* 108, 115-117.

COOMANS, A. (1979). The anterior sensilla of nematodes. *Revue de Nématologie* 2, 259-283.

DE CONINCK, L. (1942). [The symmetry relations at the anterior end of (free-living) nematodes.] *Natuurwetenchappelijk Tijdschrift* 24, 29-68.

DE CONINCK, L. (1965). Némathelminthes (Nematodes). In: Grassé, P.P. (Ed.). *Traité de zoologie, Tome IV, Fasc. 2*. Paris, France, Masson et Cie.

DECRAEMER, W. & COOMANS, A. (2007). Revision of some species of the genus *Paralongidorus sensu* Siddiqi *et al.* (1993), with a discussion on the relationships within the family Longidoridae (Nematoda: Dorylaimida). *Nematology* 9, 643-662. DOI: 10.1163/156854107782024776

DECRAEMER, W., COOMANS, A. & BALDWIN, J. (2014). 1. Morphology of Nematoda. In: Schmidt-Rhaesa, A. (Ed.). *Handbook of zoology. Gastrotricha, Cycloneuralia and Gnathifera. Volume 2: Nematoda*. Berlin, Germany, De Gruyter, pp. 1-59.

DE LEY, P., LOOF, P.A.A. & COOMANS, A. (1993). Terrestrial nematodes from the Galápagos Archipelago. II. Redescription of *Aporcelaimellus obtusicaudatus* (Bastian, 1865) Altherr, 1968, with review of similar species and a nomenclature for the vagina in Dorylaimida (Nematoda). *Bulletin de l'Institut Royal des Sciences Naturelles de Belgique* 63, 13-34.

HEYNS, J. (1965). On the morphology and taxonomy of the Aporcelaimidae, a new family of dorylaimid nematodes. *Entomology Memoirs, Department of Agriculture Technical Services, Republic of South Africa* 10, 1-51.

PEÑA-SANTIAGO, R. & ABOLAFIA, J. (2007). Notes on the genus *Ecumenicus* Thorne, 1974 (Dorylaimida: Qudsianematidae). *Journal of Nematode Morphology and Systematics* 10, 97-100.

Chapter 5

Stoma and feeding apparatus

In dorylaims, the stoma is a tube-like structure connecting the oral aperture to the pharyngeal lining, and appears to be armed with a protrusible structure. It consists of three major regions or sections, namely vestibulum, guiding sheath (with the protrusible structure inside) and odontophore. Coomans (1964) tried to homologise the dorylaimid stoma with that of other nematodes and extrapolated the nomenclature proposed by Steiner (1933) for the stomatal sections in the Rhabditida (cheilo-, pro-, meso-, meta- and telostom) to the dorylaims. With due caution (*cf.* Jairajpuri & Ahmad, 1992), such action is herein followed for practical purposes since it at least provides a good starting point to explain the complex organisation of this important element of dorylaimid anatomy and functional morphology. Lippens (1975) presented a very detailed compendium of the anatomy and histology of dorylaimid stoma.

Cheilostom

Detailed structure

The anteriormost portion of the stoma is termed the cheilostom or vestibulum – some authors refer to this structure simply as the stoma – and it runs from the oral aperture to the guiding ring. As derived from TEM studies of *Aporcelaimellus obtusicaudatus*, its detailed structure can be summarised as follows (Lippens *et al.*, 1974; Lippens, 1975; see Fig. 5.1). The stomatal aperture leads into the cheilostom, the lining of which is in fact an extension of the labial cuticle, where the cortical layer is reduced and the granular layer strongly developed. At the posterior end of the cheilostom, its lining connects with the lining of the guiding sheath (see below). This junction acts as a guiding ring, but a differentiated sclerotised ring does not exist. The basal layer of the cheilostomatal lining projects beyond the guiding ring as a cuticular cylinder. A thin layer of epidermis surrounds the cuticular cylinder at both its inner and

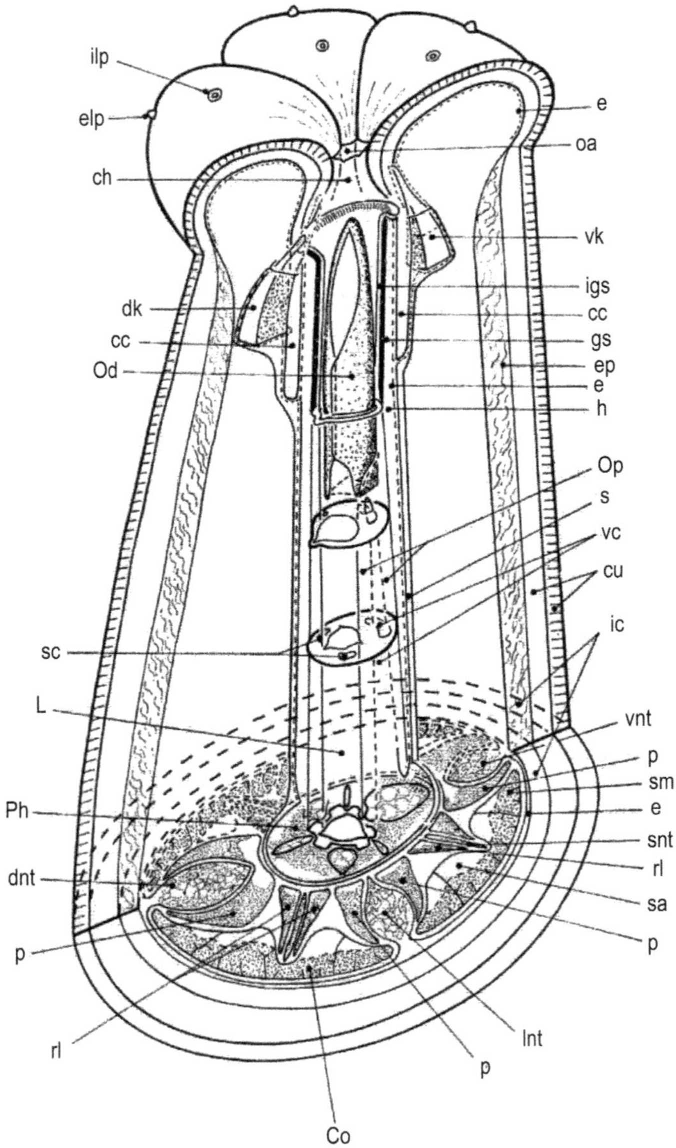
ilp
elp
ch
dk
cc
Od
sc
L
Ph
dnt
p
rl
e
oa
vk
igs
cc
gs
ep
e
h
Op
s
vc
cu
ic
vnt
p
sm
e
snt
rl
sa
p
lnt
p
Co

outer side, being in contact with the anteriormost part of the guiding sheath.

General morphology

As observed with LM, the cheilostom is sometimes nearly cylindrical (Fig. 5.2A, D), but more often it resembles a truncate cone, slightly tapering at its junction with the labial cuticle (Fig. 5.2B, E). Its wall may be thin and difficult to distinguish (Fig. 5.2A), but in some cases it appears thicker and is easily seen (Fig. 5.2B, D, E), occasionally becoming sclerotised. Very exceptionally, the cheilostom displays other differentiations. Thus, it is flask-like (Fig. 5.2C) and slightly refractive in members of the genus *Proleptonchus*. The representatives of the family Aulolaimoididae are distinguishable by having rib-like elements around the cheilostom, this appearing as a basket-like structure (Fig. 5.2F). Its anterior extremity, surrounding the oral aperture, becomes visibly thickened in many representatives of the family Tylencholaimidae (Fig. 5.2G) and is provided with four circumoral refractive platelets in *Pungentus* species (Fig. 5.2H, I).

Of special mention is the greatly modified, strongly sclerotised, chamber-like cheilostom observed in representatives of the genus *Aetholaimus* (Nygolaimidae) (Fig. 5.3A), the subfamily Carcharolaiminae (Qudsianematidae) (Fig. 5.3B) and the family Actinolaimidae (Fig. 5.3C). In all these cases the cheilostom invariably consists of two labial and post-labial sections – an excellent example of parallel evolution in the Dorylaimida. The labial portion becomes a large, roomy cavity occupying most of the lip region diameter, whereas the postlabial portion

Fig. 5.1. *Three-dimensional reconstruction to scale of the anterior part of the feeding apparatus of* Aporcelaimellus obtusicaudatus. *Abbreviations: cc = cuticular cylinder; ch = cheilostom; Co = contractile part; cu = cuticle; dk = dorsal crest; dnt = dorsal nerve trunk; e = epidermis; elp = external labial papillae; ep = epidermal projection; gs = guiding sheath; h = 'hydrostatic tissue'; ic = infracuticle; igs = inner lining of guiding sheath; ilp = inner labial papillae; L = lumen; lnt = lateral nerve trunk; Od = odontostyle; oa = oral aperture; Op = odontophore; Ph = pharynx; p = protractor; rl = anterior retractor; s = stomodeal membrane; sa = sarcoplasm; sc = subdorsal core; sm = somatic musculature; snt = submedian nerve trunk; vc = ventral core; vk = ventral crest; vnt = ventral nerve trunk. (After Lippens* et al. *(1974),* Nematologica *20, p. 247, Fig. 29.)*

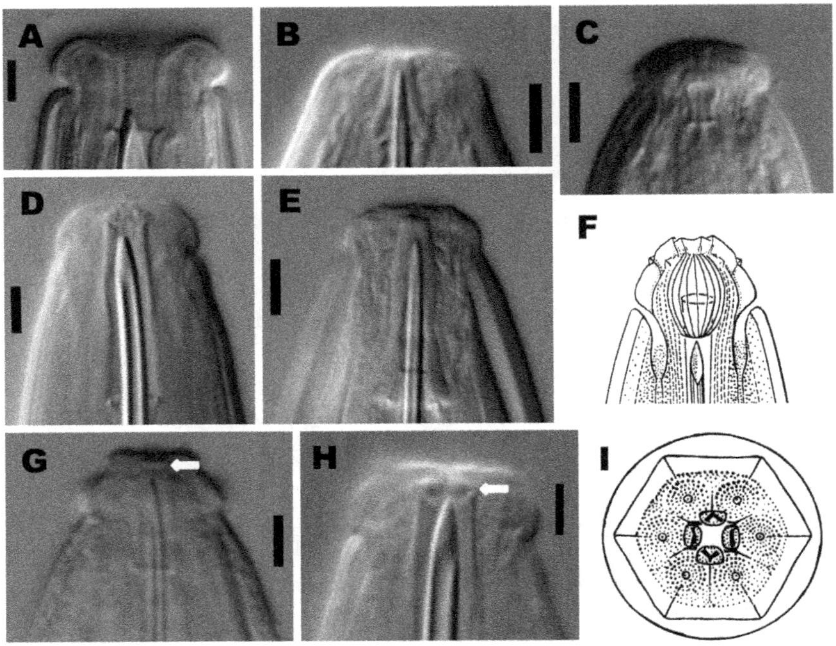

Fig. 5.2. *Morphology of the cheilostom (part I). A, D: Cylindrical; B, E: A truncate cone; C: Flask-like; F: Basket-like; G-I: Bearing perioral platelets (arrow). (A:* Aporcelaimellus communis *from Spain. B:* Rhyssocolpus iuventutis *from Spain. C:* Proleptonchus *sp. from Spain. D:* Pungentus silvestris *from Spain. E:* Enchodelus longispiculus *from Iran. F:* Aulolaimoides elegans, *after Thorne (1939),* Capita Zoologica *8, p. 247, Plate XXVIII-Fig. 205a. G:* Chitwoodiellus heynsi *from Spain. H:* Pungentus silvestris *from Spain. I:* Pungentus pungens, *after Thorne & Swanger (1936),* Capita Zoologica *6, p. 217, Plate XXX-Fig. 180a.) (Scale bars = 5 μm.)*

is a nearly as long as wide chamber. The labial portion of the cheilostom in *Aetholaimus*, as described by Coomans & Loof (1978), ends anteriorly in curved ridges and at its base with six well-developed radial differentiations, each opposite the centre of a lip and having two lateral projections, whereas the postlabial section consists of six elongate triangular to oval, interlabial sclerotisations in its anterior half, and narrow walls in its posterior half. In *Carcharolaimus* species, the labial portion of the cheilostom is a basket-like structure made of six pairs of curved plates – appearing as ribs in lateral view (Fig. 5.3B$_1$-B$_2$) – that anteriorly bear fine transverse striations and posteriorly numerous warts and ridges

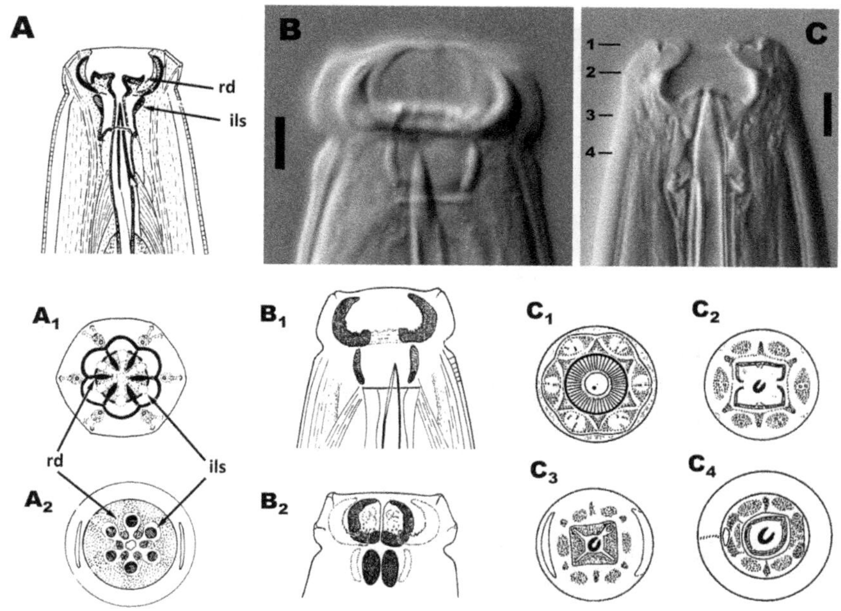

Fig. 5.3. *Morphology of the cheilostom (part II). A:* Aetholaimus rotundicauda*;* A_1 *= cross section at level of labial section;* A_2 *= cross section at level of postlabial section, rd = radial differentiations; ils = interlabial sclerotisations (after Coomans & Loof (1978),* Proceedings of the Helminthological Society of Washington *45, p. 89, Fig. 4E, H, I.); B:* Carcharolaimus eximius*;* B_1 *= lateral median view,* B_2 *= lateral submedian view (after Peña-Santiago & Liébanas (1994),* Fundamental and Applied Nematology *17, p. 408, Fig. 2A, B). C:* Paractinolaimus *sp. from Iberian Peninsula;* C_1*-*C_4 *= cross-sections at level of 1-4 in* Parastomachoglossa costata *(=* Actinca costata*), respectively (after Coomans & Loof (1986),* Revue de Nématologie *9, p. 224, Fig. 2A-D.) (Scale bars = 5 µm.)*

(denticle-like structures). The basal aperture of the basket is surrounded by six platelets in a radial arrangement, whereas the postlabial section is made up of six pillars thickened both anteriorly and posteriorly, apparently hollow and posteriorly fused with the guiding ring (Peña-Santiago & Liébanas, 1994; De Ley *et al.*, 2005). As observed by Coomans & Loof (1986) in *Parastomachoglossa costata* (= *Actinca costata*) (Fig. 5.3C_1-C_4), an actinolaimid species, the oral aperture is surrounded by sclerotised rings that are interconnected by radial spoke-like structures (Fig. 5.3C_1), the labial part of the cheilostom becoming quadrangular

(Fig. 5.3C_2), and the postlabial section developing four (one dorsal, one ventral and two lateral; Fig. 5.3C_3) broad tooth-like structures (onchia), and ending as a hexaradiate structure.

Guiding ring

As mentioned above, the posterior end of the cheilostom joins the extreme front end of the guiding sheath, originating as a differentiation that serves as a guide for odontostyle movement (see below) and therefore termed the 'guiding ring' (Fig. 5.4). It is a delicate structure, especially distinct in well-preserved specimens. Most dorylaims possess a simple guiding ring (Fig. 5.4A, D), consisting of a fixed, more or less sclerotised, annular structure. Several genera, however, bear a double guiding ring (Fig. 5.4B, E) since, in addition to the fixed ring, there is a second, less sclerotised, annulus formed at the top of a short projection of the fixed ring into the cheilostom lumen. In other cases, the guiding ring is weakly sclerotised, thin and with an irregular layout, which is known as a plicate guiding ring (Fig. 5.4C, F). The nature and location (its distance to the anterior body end) of the guiding ring are useful traits for taxonomical purposes.

Taylor & Robertson (1971) studied in great detail the nature of the guiding ring in longidorid species and found a remarkable variability in its nature. In the genus *Xiphinema* it is a structure situated at the level of the base of the cuticular wall of the stoma and separated from this in the form of a distinct annulus with an electron-dense centre and connected with the guiding sheath by means of a thin and flexible extension (Fig 5.4G). In *Longidorus*, however, it is an annulus of granular material, more electron-dense than in *Xiphinema*, also attached to the guiding sheath by a short, flexible connection yet remarkable because of the presence of an amorphous, apparently fluid, tissue between the guiding sheath and the basement membrane of the stoma (Fig. 5.4H).

Guiding sheath

The intermediate section of the stoma is the guiding sheath. It runs from the guiding ring until its connection with the anterior end of the odontophore. Three stomatal subregions (pro-, meso- and metastom) form part of this section, which bears the protrusible structure

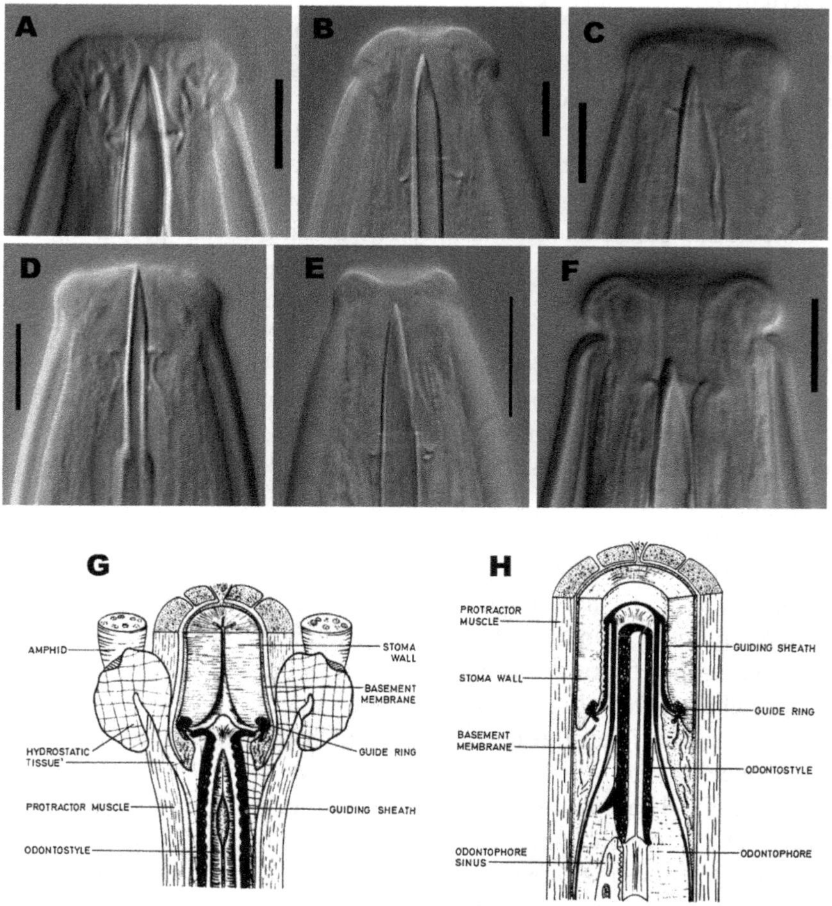

Fig. 5.4. *Guiding ring. A, D: Simple (single); B, E: Double; C, F: Plicate; G: Diagram of anterior part of odontostyle and guiding sheath apparatus in* Longidorus elongatus *(after Taylor & Robertson (1971),* Nematologica *17, p. 305, Fig. 2); H: Diagram of anterior part of odontostyle and guiding apparatus in* Xiphinema diversicaudatum *(after Taylor & Robertson (1971),* Nematologica *17, p. 304, Fig. 1). (A:* Crassolabium persicum *from Iran. B:* Nevadanema nevadense *from Spain. C:* Aporcella simplex *from Spain. D:* Eudorylaimus *sp. from Spain. E:* Dorylaimus *sp. from Spain. F:* Aporcelaimellus communis *from Spain.) (Scale bars: A-D, F = 10 μm; E = 20 μm.)*

(either a mural tooth or an axial odontostyle) that characterises the dorylaims.

As revealed by TEM in free-living forms (Lippens *et al.*, 1974; Lippens, 1975; Grootaert & Wyss, 1978), the guiding sheath is electron-dense, centrally covered by a relatively thin layer of low electron density. The following parts (layers) can be seen at a level a little posterior to the guiding ring from the lumen towards the periphery (Fig. 5.1): inner lining, guiding sheath proper, 'hydrostatic tissue', epidermis, cuticular cylinder, and stomodeal membrane. The guiding sheath proper appears as packed granules and fibres, and is surrounded by what has been called 'hydrostatic tissue' (Taylor & Robertson, 1971) or 'hydrostatic substance' (Coomans, 1975; Robertson & Taylor, 1975), the latter with the same composition of granules and fibres. Around the 'hydrostatic tissue' a sheath of epidermis is present, continuous with that surrounding the cheilostom cuticle. The stomodeal membrane enclosing the pharynx continues anteriad alongside the epidermal sheath to near the guiding ring.

Light microscopy offers a much more simplified morphology of the guiding sheath. Leaving aside the presence of the enclosed protrusible structure, it is a short region of the stoma that lacks any relevant differentiation. Coomans (1964) explained in detail this structure and his interpretation is followed here. In members of the suborder Nygolaimina it is double-walled (Fig. 5.5A) since the cuticular lining has probably been separated from the tissue from which it was secreted so that a cavity occurs, with the internal core (cuticular lining) thicker than the surrounding sheath and often plicate. Posteriorly, in the left subventral sector of the median part, a mural tooth (see below) is inserted on a support. The tooth is movable in a longitudinal direction; if protruded, the stomatal lining accompanies it, thus forming a guiding sheath that, together with the guiding ring, constitutes the guiding apparatus (Fig. 5.5C, D). By contrast, in members of the suborder Dorylaimina the guiding sheath is a tube-like structure (Fig. 5.4B) whose wall is less developed than in Nygolaimina but provides a relatively broader support for the odontostyle than that provided for the mural tooth in Nygolaimina. When the odontostyle is completely withdrawn, the guiding sheath is stretched and often shows no folds anterior to the guiding ring, whereas if the odontostyle is protruded, the guiding sheath accompanies it and forms a fold inside the fixed guiding ring (Fig. 5.5E, F).

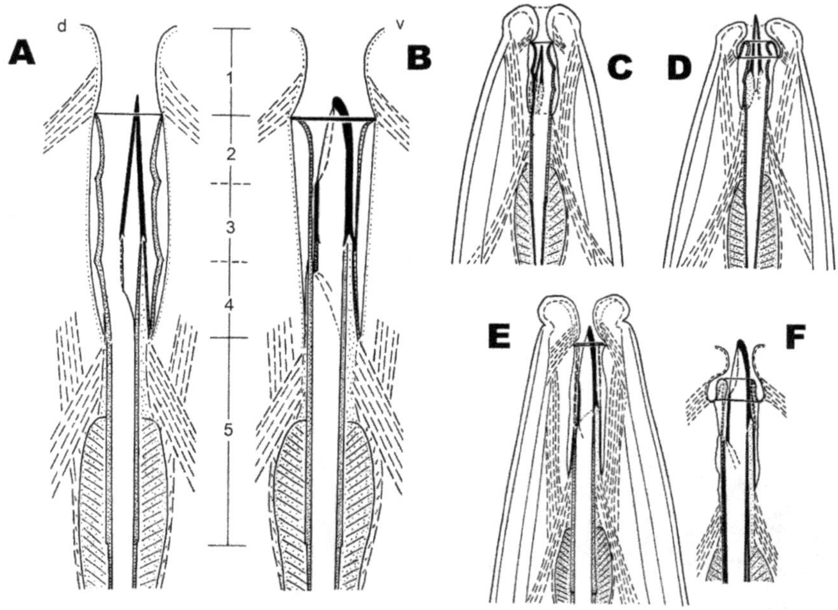

Fig. 5.5. *The stoma of dorylaims. A: Nygolaimina pattern; B: Dorylaimina pattern; C: Guiding sheath of Nygolaimina with re-drawn mural tooth; D: Same with protruded mural tooth; E: Guiding sheath of Dorylaimina with re-drawn odontostyle; F: Same with protruded odontostyle. 1 = cheilostom, 2 = prostom, 3 = mesostom, 4 = metastom, 5 = telostom, d = dorsal, v = ventral. (After Coomans (1964),* Nematologica *9(1963), pp. 588, 590, 592; Figs 1C, D; 2; 3A, B.)*

Mural tooth

In members of the suborder Nygolaimina the guiding sheath is armed with a mural tooth. It is a protrusible structure attached to the stomatal wall at its subventral median side on a support and lacks a distinct aperture. Its form exhibits some variation (Fig. 5.6A), which was very well described by Heyns (1968) in his fundamental monograph devoted to this dorylaimid taxon. It may be deltoid, *i.e.*, conical and with a distinct lumen (Fig. 5.6A1; B); solidideltoid, with its anterior part lacking a distinct lumen which the posterior one possesses (Fig. 5.6A2); acicular, needle-like, very thin and without perceptible lumen (Fig. 5.6A3); dorylaimoid, tube-like and with a sloping anterior portion (Fig. 5.6A4-5; C) or linear with parallel walls and a narrow but visible

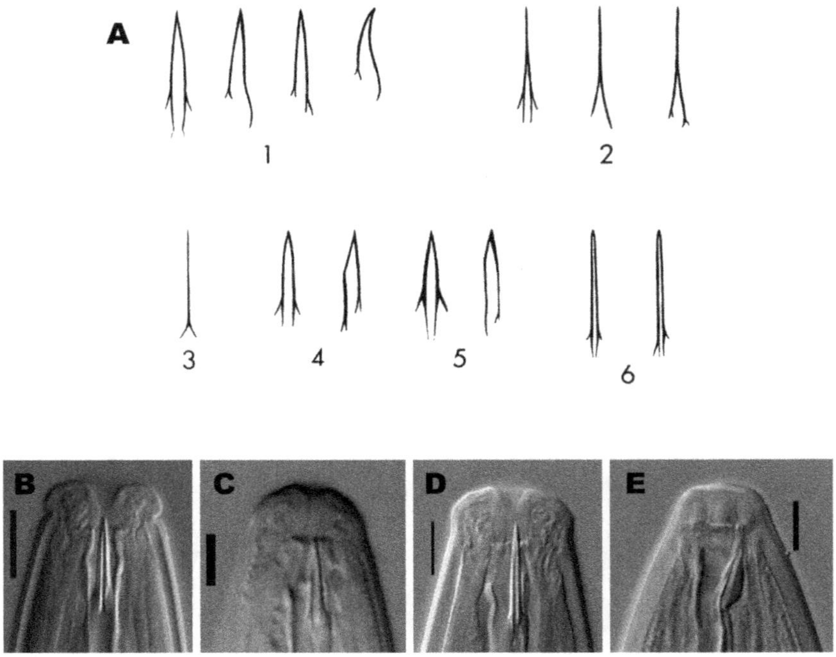

Fig. 5.6. *Mural tooth in the Nygolaimina. A: Deltoid (1), solidideltoid (2), acicular (3), dorylaimoid (4, 5) and linear (6) (diagrammatic; after Heyns (1968),* Entomology Memoirs, Department of Agricultural Technical Services, Republic of South Africa *19, p. 95, Figs 1-6.); B: Deltoid mural tooth of* Nygolaimus brachyuris *from Spain; C: Dorylaimoid mural tooth of* Laevides *sp. from the USA; D: Linear mural tooth of* Aquatides *sp. from Spain; E: Bent mural tooth of* Nygolaimium *sp. from Costa Rica. (Scale bars: B, D, E = 10 μm; C = 5 μm.)*

lumen (Fig. 5.6A6; D). The mural tooth is typically held vertical and straight, more or less parallel to the body axis. Occasionally, however, it appears bent and visibly curved (Fig. 5.6.E). Thorne (1930) stated that nygolaims, which are predaceous nematodes, puncture their prey by means of the tooth and afterwards the liquid food is sucked out and passes 'along' the tooth (which meanwhile had been withdrawn) (*cf.* Coomans, 1964).

Axial odontostyle

Most dorylaims, the members of the suborder Dorylaimina, bear another kind of protrusible structure inside the guiding sheath, the odontostyle[1] (Figs 5.5B; 5.7). It is typically an axial, symmetrical, tube-like structure with a sloping aperture on its dorsal side, attached to the median guiding sheath, and in continuity with the posteriormost region of the stoma (telostom or odontophore, see below), thereby forming a functional unit. When the dorylaim punctures its prey or other material, liquid food passes 'through' the odontostyle, which remains protruded during the feeding act (*cf.* Linford, 1937; Coomans, 1964). As observed under SEM (Lippens *et al.*, 1974), the odontostyle of *Aporcelaimellus obtusicaudatus* has a solid tip, a wide aperture whose enclosing rims become swollen, and a narrow longitudinal slit along the dorsal side, with one of the rims overlapping the other somewhat, and which widens progressively towards the base (Fig. 5.1).

The morphology of the odontostyle is extremely variable among dorylaimid taxa (Fig. 5.7) and has high taxonomic value for species identification, varying in morphometrics and other characters. The absolute length usually ranges from 10 to 30 μm, but smaller (only a few micrometres long) and much larger sizes (exceptionally over 200 μm in some longidorid forms) also occur. Its relative morphometrics (ratios) are of some interest. Thus, odontostyle length divided by lip region diameter ratio is very constant within the same species, but highly variable when species are compared, with forms bearing relatively short (ratio visibly less than 1.0; Fig. 5.7G, H, M), medium-sized (ratio around 1.0; Fig. 5.7C-F) or long (ratio appreciably more than 1.0; Fig. 5.7I-K) odontostyle. On the other hand, the odontostyle length to width ratio provides a first approximation to the general shape or appearance of the odontostyle. For example, many dorylaims bear a strong or robust odontostyle, up to 6-7 times as long as wide, with a very distinct lumen and aperture (Fig. 5.7A-F), whereas other forms possess an attenuate odontostyle, often more than 6-7 times as long as wide, with a narrow lumen and small aperture (Fig. 5.7H-J, M), which in particular taxa

[1] Many authors, especially in old contributions, used the terms 'stylet' or 'spear' to describe or to refer to the protrusible structure of Dorylaimina as a synonym either of the odontostyle or the functional unit constituted by the odontostyle plus the odontophore. However, the former terms are now obsolete.

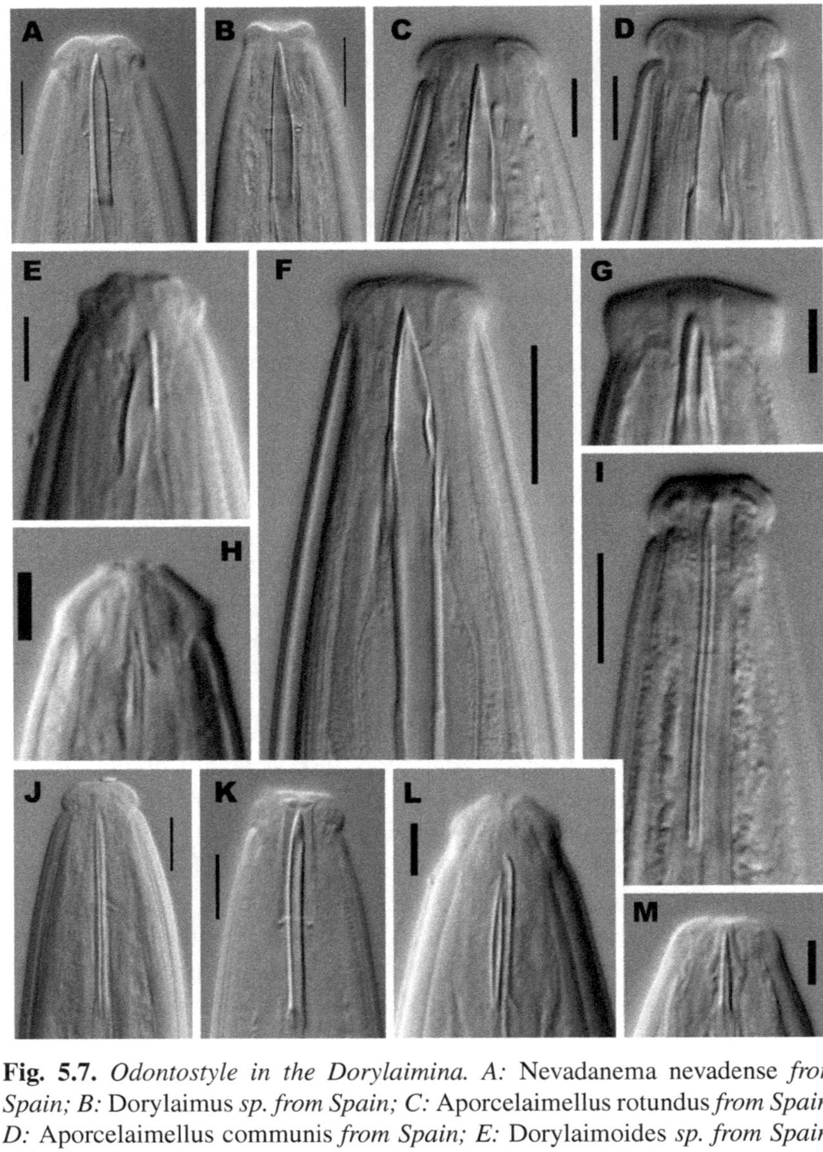

Fig. 5.7. *Odontostyle in the Dorylaimina. A:* Nevadanema nevadense *from Spain; B:* Dorylaimus *sp. from Spain; C:* Aporcelaimellus rotundus *from Spain; D:* Aporcelaimellus communis *from Spain; E:* Dorylaimoides *sp. from Spain; F:* Aporcella simplex *from Spain; G:* Pachydorylaimus notabenus *from Costa Rica; H:* Funaria *sp. from Spain; I:* Longidorella deliblatica *from Spain; J:* Heterodorus geraldi *from Romania; K:* Pungentus silvestris *from Spain; L:* Tylencholaimellus *sp. from Spain; M:* Rhyssocolpus iuventutis *from Spain. (Scale bars: A, B, F = 20 μm; C, D, I-K = 10 μm; E, G, H, L, M = 5 μm.)*

(longidorid and some nordiid representatives) are so extreme that the odontostyle becomes a long, needle-like structure. The length of the odontostyle aperture is a remarkable trait, especially in forms with a strong odontostyle, the aperture ranging from less than one-fifth to more than one-half of the total odontostyle length, as shown in Fig. 5.7A-D in which the odontostyle aperture increases in length from A to D.

Leaving aside its morphometrics, the odontostyle may display many other relevant modifications or specialisations of taxonomic interest. Some examples are discussed hereunder. The odontostyle is usually straight, but sometimes becomes more or less visibly arcuate (Fig. 5.7H, K), in particular in those taxa bearing a long and/or attenuate odontostyle. The usual symmetry of the odontostyle is occasionally modified to produce more irregular shapes, in which the odontostyle becomes sigmoid or falcate, has its dorsal arm much longer than the ventral one (Fig. 5.7E), *etc*. In some cases, its wall appears especially thick, forming a massive structure (Fig. 5.7G), often with a narrow lumen, small aperture and visibly bifurcate roots. In some members of the family Tylencholaimellidae the odontostyle bears a stiffening dorsal (Fig. 5.7L) or ventral piece.

The stomatal protrusible structure observed in some members of the family Aporcelaimidae deserves special mention, for example, the representatives of the genera *Aporcelaimoides* and *Sectonema*, in which a reduced odontostyle (Fig. 5.8A), a mural tooth (Fig. 5.8B) or even a mural odontostyle (Fig. 5.8C) may be present. The general aspect of these structures, in particular the mural tooth and the mural odontostyle, resembles that found in nygolaims – it probably represents another example of evolutionary convergence in the Dorylaimida – but, according to Heyns (1965), it should be interpreted as the culmination of a progressive reduction of odontostyle size as well as an increase in its aperture size (Fig. 5.8D).

Odontophore (telostom)

The posteriormost section of the stoma is termed the odontophore and, with due caution (see above), might be homologous to the telostom of other nematode taxa. Anteriorly it joins the guiding sheath, forming a functional unit with the odontostyle, and posteriorly connects with the pharyngeal lining. Its posterior part is surrounded by pharyngeal tissue

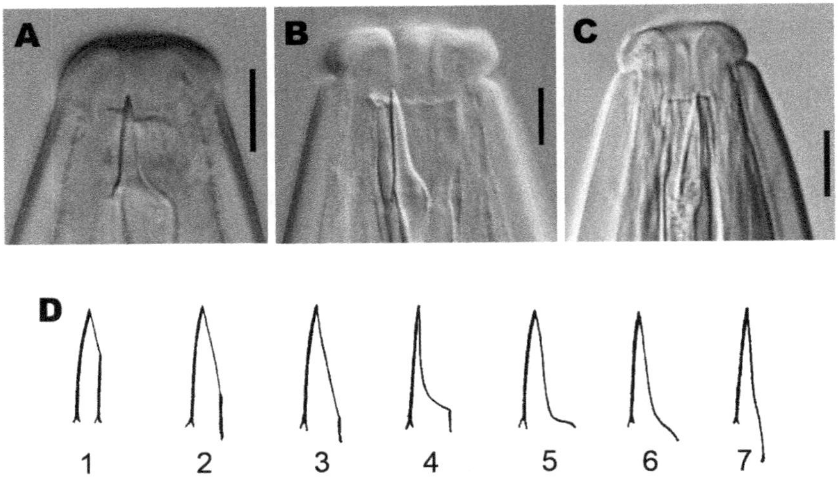

Fig. 5.8. *The modified odontostyle observed in members of the family Aporcelaimidae. A: Reduced odontostyle of* Sectonema vietnamense *from Vietnam; B: Mural tooth of* Sectonema demani *from Belgium; C: Mural odontostyle of* Aporcelaimoides brevistylus *from Vietnam; D: Proposed evolution of the odontostyle aperture in members of Aporcelaimidae (after Heyns (1965),* Entomology Memoirs, Department of Agricultural Technical Services, Republic of South Africa *10, p. 33, Figs 1-7). 1 = Typical dorylaimid odontostyle; 2 =* Aporcelaimus americanus; *3 =* A. regius*; 4 =* Epacrolaimus vorax*; 5 =* Sectonema ventrale*; 6 =* S. macrospiculum*; 7 =* S. basilgoodeyi.*) (Scale bars = 10 μm.)*

(Figs 5.5A, B; 5.7F; 5.9). As revealed by TEM (Lippens *et al.*, 1974; Lippens, 1975; Fig. 5.1), the odontophore consists of one ventral and two subventral sectors, the former being much larger than the latter. The ventral side of the odontostyle connects with the odontophore, the ventral tissue core in the odontophore reaching anteriad nearly to the point of insertion of the odontostyle, whereas the subdorsal cores do not reach that far. The cores open into the pharynx and from there only the central part of the stomodeal lining (= basal part of the odontophore) proceeds posteriorly through the pharyngeal tissues. The anterior portion of the odontophore is surrounded by hydrostatic tissue, epidermis and stomodeal membrane, while the posterior portion, located within the pharynx, is surrounded by a very thin layer of epidermis with several lacunae occurring throughout its length. A differentiation similar to the platelet structure observed in the pharyngeal lining

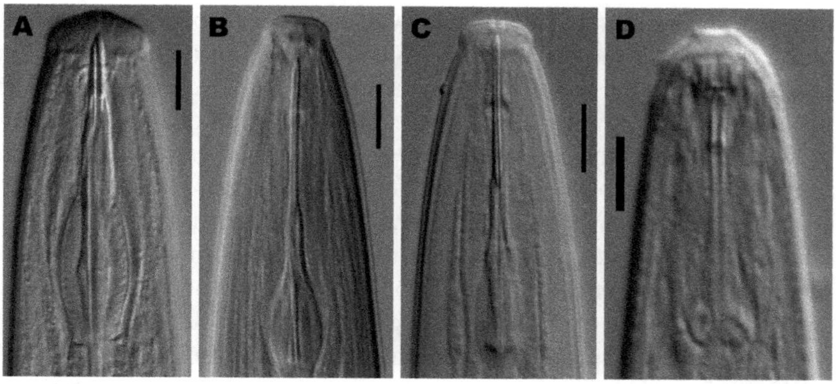

Fig. 5.9. *Odontophore in the Dorylaimina. A:* Pachydorylaimus holovachovi *from Costa Rica; B:* Enchodelus carpaticus *from Romania; C:* E. saxifragae *from Romania; D:* Oostenbrinkia *sp. from Iran. (Scale bars: A = 10 μm; B, C = 20 μm; D = 5 μm.)*

becomes progressively more obvious as the junction to the pharynx is approached. Although structurally the odontophore and pharyngeal lining are continuous, the actual transition is marked by a difference in thickness.

As observed under LM, the odontophore is usually a tube-like structure, with linear walls and no special modification except for a weak thickening at its junction with the pharyngeal lining (Figs 5.5A, B; 5.7F). Nevertheless, it may display some specialisations in particular taxa. Occasionally, its walls are visibly robust throughout its length (Fig. 5.9A) and it is visibly arcuate (curved) and/or refractive in representatives of the family Mydonomidae. In members of several genera, for example, *Enchodelus* and *Xiphinema*, its posterior half is noticeably thickened, bearing three flange-like extensions (Fig. 5.9B, C). Its posterior extremity (base) has more or less developed knobs (Fig. 5.9D) in several genera such as *Tylencholaimellus*, *Tylencholaimus* and others.

Musculature associated with the stoma

According to Coomans (1964; see also Coomans & Loof, 1986), three sets of specialised, stomatal muscles are distinguishable in the Dorylaimida (Fig. 5.5): *i*) the *dilatatores buccae* extend anteriorly from the subdorsal and subventral body wall to the wall of the cheilostom in the vicinity of the guiding ring. When the wall of the cheilostom is folded, this may give the impression that these muscles are attached posterior to the guiding ring. These muscles are probably necessary to widen the cheilostom before the odontostyle is protruded; *ii*) the 'retractor muscles of the stylet' are four muscles that extend anteriorly from the submedian (laterodorsal and lateroventral) body wall to the stoma wall at about the junction of the median and proximal parts of the stoma. Each retractor consists of two muscle bands; and *iii*) the 'protractor muscles of the stylet' are eight muscles (four sublateral, two subdorsal and two subventral) extending posteriorly from the body wall to the region of the anterior pharyngeal swelling. The protractor and retractor muscles are probably antagonistic in action.

Detailed TEM studies of free-living forms (Lippens *et al.*, 1974; Lippens, 1975; Grootaert & Wyss, 1978) and longidorid species (Wright, 1965; Aboul-Eid, 1969; Wright *et al.*, 1983, Jacobs *et al.*, 1987) have confirmed the observations made with LM. Thus, there are eight protractor muscles (two subdorsal, two laterodorsal, two lateroventral and two subventral) running from the thick stomatal cuticle anterior to the guiding ring, where they form one dorsal and one ventral group, until the base of the odontophore. In general, the fine structure of these muscles resembles that of somatic muscles.

References

ABOUL-EID, H.Z. (1969). Electron microscope studies on the body wall and feeding apparatus of *Longidorus macrosoma*. *Nematologica* 15, 451-463. DOI: 10.1163/187529269X00786

COOMANS, A. (1964). Stoma structure in members of the Dorylaimina. *Nematologica* 9 (1963), 587-601. DOI: 10.1163/187529263X00692

COOMANS, A. (1975). Morphology of Longidoridae. In: Lamberti, F., Taylor, C.E. & Seinhorst, J.W. (Eds). *Nematode vectors of plant viruses*. London, UK, Plenum Press, pp. 15-37.

COOMANS, A. & LOOF, P.A.A. (1978). Observations on the subfamily Aetholaimidae Jairajpuri, 1965 (Nygolaimidae: Nematoda). *Proceedings of the Helminthological Society of Washington* 45, 82-92.

COOMANS, A. & LOOF, P.A.A. (1986). Redescription of *Actinolaimus costatus* Schneider, 1935 (Nematoda: Actinolaimidae), with observations on its pharyngosomatic muscles and discussion of its taxonomic position. *Revue de Nématologie* 9, 221-231.

GROOTAERT, P. & WYSS, U. (1978). Ultrastructure and function of the anterior feeding apparatus in *Labronema* sp. *Nematologica* 24, 243-250. DOI: 10.1163/187529278X00209

DE LEY, P., VAN DRIESSCHE, R. & COOMANS, A. (2005). Terrestrial nematodes from the Galápagos Archipelago. 10. Morphological and taxonomical analysis of *Carcharolaimus ramirezi* Thorne, 1967 (Dorylaimida: Qudsianematidae). *Nematology* 7, 133-147. DOI: 10.1163/1568541054192135

HEYNS, J. (1965). On the morphology and taxonomy of the Aporcelaimidae, a new family of dorylaimid nematodes. *Entomology Memoirs, Department of Agriculture Technical Services, Republic of South Africa* 10, 1-51.

HEYNS, J. (1968). A monographic study of the nematode families Nygolaimidae and Nygolaimellidae. *Entomology Memoirs, Department of Agricultural Technical Services, Republic of South Africa* 19, 1-144.

JACOBS, P.J.F., HEYNS, J., COOMANS, A. & LOOTS, G.C. (1987). Studies on the ultrastructure of the pharyngeal region in Longidoridae (Nematoda). *Nematologica* 33, 199-212. DOI: 10.1163/187529287X00335

JAIRAJPURI, M.S. & AHMAD, W. (1992). *Dorylaimida. Free-living, predaceous and plant-parasitic nematodes.* New Delhi, India, Oxford & IBH Publishing Co. Pvt. Ltd.

LINFORD, M.B. (1937). The feeding of some hollow-stylet nematodes. *Proceedings of the Helminthological Society of Washington* 4, 41-46.

LIPPENS, P.L. (1975). Ultrastructure of Dorylaimidae. In: Lamberti, F., Taylor, C.E. & Seinhorst, J.W. (Eds). *Nematode vectors of plant viruses.* London, UK, Plenum Press, pp. 159-178.

LIPPENS, P.L., COOMANS, A., DE GRISSE, A.T. & LAGASSE, A. (1974). Ultrastructure of the anterior body region in *Aporcelaimellus obtusicaudatus* and *A. obscurus. Nematologica* 20, 242-256. DOI: 10.1163/187529274X00212

PEÑA-SANTIAGO, R. & LIÉBANAS, G. (1994). Nematodes of the order Dorylaimida from Andalucía Oriental, Spain. The genus *Carcharolaimus* Thorne, 1939 with description of *C. eximius* sp. n. and a compendium of the genus. *Fundamental and Applied Nematology* 17, 405-413.

ROBERTSON, W.M. & TAYLOR, C.E. (1975). The structure and musculature of the feeding apparatus in *Longidorus* and *Xiphinema*. In: Lamberti, F., Taylor, C.E. & Seinhorst, J.W. (Eds). *Nematode vectors of plant viruses.* London, UK, Plenum Press, pp. 179-194.

STEINER, G. (1933). The nematode *Cylindrogaster longistoma* (Stefanski) Goodey, and its relationship. *Journal of Parasitology* 20, 66-68.

TAYLOR, C.E. & ROBERTSON, W.M. (1971). Ultrastructure of the guide ring and guiding sheath in *Xiphinema* and *Longidorus. Nematologica* 17, 303-307. DOI: 10.1163/187529271X00152

THORNE, G. (1930). Predacious nemas of the genus *Nygolaimus* and a new genus, *Sectonema. Journal of Agriculture Research* 41, 445-466.

THORNE, G. (1939). A monograph of the nematodes of the superfamily Dorylaimoidea. *Capita Zoologica* 8, 1-261.

THORNE, G. & SWANGER, H.H. (1936). A monograph of the nematode genera *Dorylaimus* Dujardin, *Aporcelaimus* n. gen., *Dorylaimoides* n. gen., and *Pungentus* n. gen. *Capita Zoologica* 6, 1-223.

WRIGHT, K.A. (1965). The histology of the oesophageal region of *Xiphinema index* Thorne and Allen, 1950, as seen with electron microscope. *Canadian Journal of Zoology* 43, 689-700.

WRIGHT, K.A., CARTER, R.F. & ROBERTSON, W.M. (1983). The musculature of the anterior feeding apparatus of *Xiphinema* species (Nematoda: Dorylaimoidea). *Nematologica* 29, 49-64. DOI: 10.1163/187529283X00186

Nematology Monographs & Perspectives, 2021, Vol. 13, 59-80

Chapter 6

Digestive tract

The digestive tract is a tube-like structure that invariably consists of five regional specialisations: pharynx, pharyngo-intestinal junction (cardia), intestine, prerectum and rectum.

Pharynx[1]

General morphology

One of the most characteristic and distinctive traits of dorylaims is the morphology of their pharynx. It is a 'bottle-like', muscular organ, typically consisting of a slender anterior region gradually enlarging into an expanded, posterior (basal) portion[2] (Fig. 6.1A, B). Exceptionally, the representatives of the rare family Aulolaimoididae have a tripartite pharynx, with a long slender anterior section gradually expanding in its posterior part, a narrower intermediate section enveloped by glandular tissues, and the typical basal enlargement (Fig. 6.1C, D). When observed in transverse section, and as is common in nematodes, the pharyngeal lumen appears tri-radial, with three arms or radii (two laterodorsal and one midventral) that delimit three sectors, one dorsal and two lateroventral (Fig. 6.1E, F). The basal expanded portion bears five (occasionally three) mono-nucleate glands with their respective outlets. Very occasionally, some kind of malformation affecting the pharynx has been reported for dorylaims; for example, a dorso-ventral duplication of

[1] Some authors prefer the term 'oesophagus' to refer to the pharynx, but others (for example, see Coomans *et al.*, 1978) have argued for the second option, an opinion that is herein followed.

[2] The expanded (basal) portion of the pharynx is also called the *cylindrus* by some authors (*cf.* Andrássy, 1998) because it is very often cylindrical, but this term has not been very widely used due, perhaps, to the huge variability found in the shape and the size of this region.

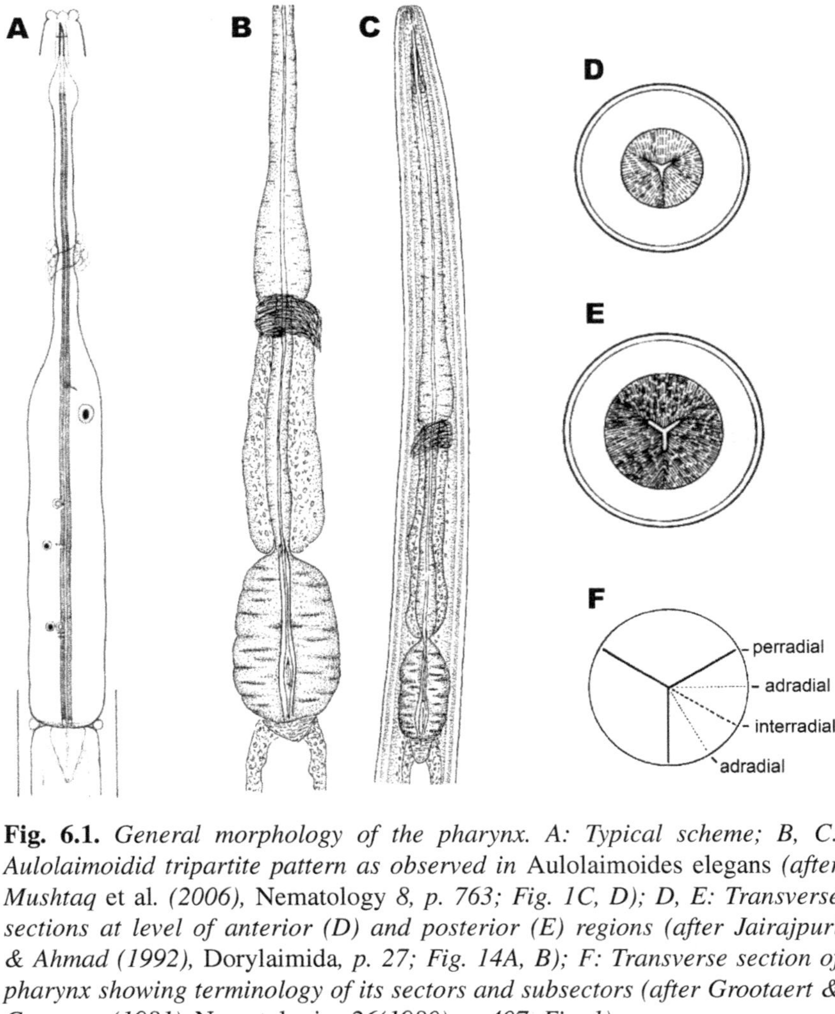

Fig. 6.1. *General morphology of the pharynx. A: Typical scheme; B, C: Aulolaimoidid tripartite pattern as observed in* Aulolaimoides elegans *(after Mushtaq* et al. *(2006),* Nematology *8, p. 763; Fig. 1C, D); D, E: Transverse sections at level of anterior (D) and posterior (E) regions (after Jairajpuri & Ahmad (1992),* Dorylaimida, *p. 27; Fig. 14A, B); F: Transverse section of pharynx showing terminology of its sectors and subsectors (after Grootaert & Coomans (1981),* Nematologica *26(1980), p. 407, Fig. 1).*

the pharyngeal part running behind the nerve ring was observed by Ali & Prabha (1971) in *Thornenema baldum*.

SECTORS

The pharynx is made up of cells extending from the cuticularised wall of the lumen to the surrounding peripheral membrane. Grootaert & Coomans (1980) proposed a revised terminology for naming the different sectors and cells of the nematode pharynx (Fig. 6.1G): the term 'perradial' refers to each of the three primary radii of the pharynx, comprising a radius of the lumen and the cell near its tip; 'interradial' designates a radius midway between two perradii (*i.e.*, a second order radius); and 'adradial' means a radius between perradius and interradius (*i.e.*, a third order radius).

ULTRASTRUCTURE

TEM studies have revealed the fine structure of both pharyngeal sections (Taylor *et al.*, 1970; Lippens, 1975). The anterior portion is a flexible tube with a lumen of approximately 750 nm diameter bounded by a 270 nm thick cuticle. Radial muscles are absent, but located a short distance posterior to the odontophore base, three radial strands of ligamentous tissue stretch from the cuticle lining of the lumen to the periphery. The cellular organisation consists mainly of elongate cells in which large nuclei, mitochondria, microtubules and convoluted membranes are prominent. Regarding the basal expansion (or bulb, see below), its lumen is circular at both extremities but triradiate for most of its length, the lining being in the form of six triangular platelets (see also Lippens *et al.*, 1974; Grootaert & Wyss, 1978). A pair of radial muscles is attached to the innermost apex of each platelet and extends to the peripheral wall of the bulb. There are also three sets of muscles that are arranged equidistantly around the circumference of the bulb throughout its length. The complementary contraction and relaxation of these sets of muscles open and close the lumen and control the volume and the turgor of the pharyngeal bulb walls.

ANTERIOR SECTION

As mentioned above, the anterior portion of the pharynx is always more slender than the posterior. Its anterior end joins and envelops the odontophore base forming a more or less developed, spindle-shaped swelling whose cuticular lining becomes nearly continuous with the odontophore walls (Fig. 6.2A-D). Despite being slender, it is usually muscular, but in specialised (longidorid, leptonchid, *etc.*) taxa it appears particularly slender and either not muscular or only weakly so, certainly differing in its histology from the posterior portion. Exceptionally, some belondirid species show a second, small, spindle-like swelling at the anterior part of the slender pharyngeal portion located a short distance

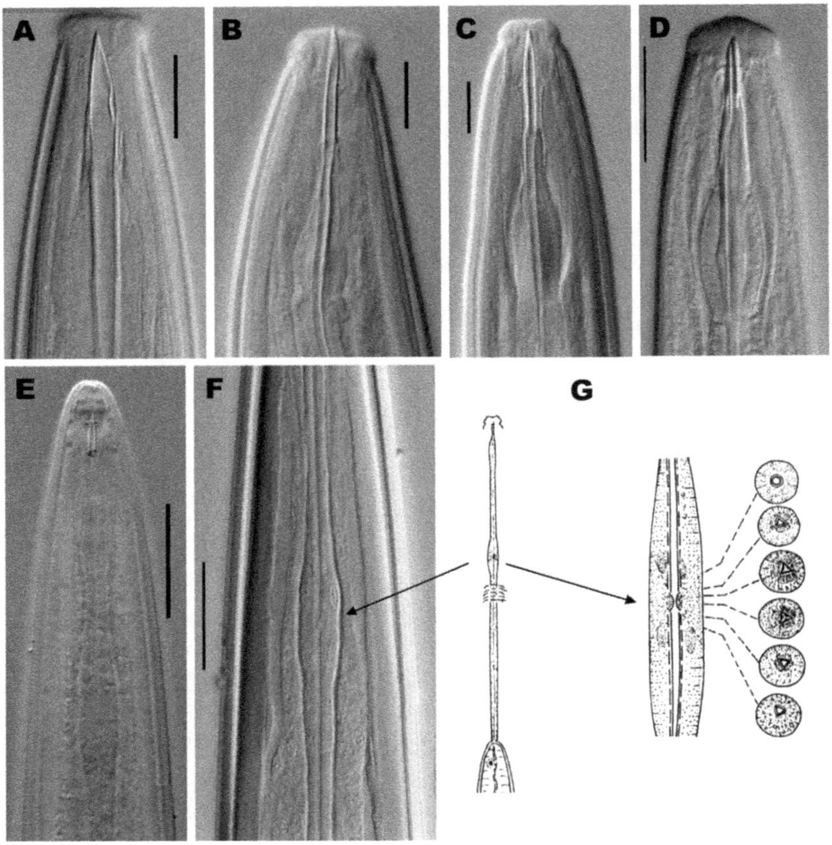

Fig. 6.2. *Slender portion of pharynx. A-D: Anterior end joining and surrounding odontophore base; E: Simple second swelling at its anterior part; F, G: Valvate swelling (after Nair & Coomans (1973),* Nematologica *18(1972), p. 500, Fig. 3C). (A:* Aporcella simplex *from Spain. B:* Dorydorella bryophila *from the USA. C:* Nevadanema nevadense *from Spain. D:* Pachydorylaimus holovachovi *from Costa Rica. E:* Belondira *sp. from Vietnam. F:* Syncheilaxonchium nairi *from Iran.) (Scale bars: A, C-F = 20 μm; B = 10 μm.)*

posterior to its junction with the odontophore base (Fig. 6.2E); very occasionally, this swelling bears valve-like sclerotisations internally (Fig. 6.2F, G). A thin sheath may surround the greater part of this pharyngeal section in *Axonchium* species (*cf.* Nair & Coomans, 1973), this probably consisting of connective tissue. Anteriorly, some muscles seem to connect the sheath with the body wall, while posteriorly, the sheath joins the muscular sheath enveloping the basal expansion. Siddiqi (1971) reported the existence of endolids, regarded as probable sensory structures, in the subventral sectors of the cuticular lining of the pharynx a short distance anterior to the nerve ring in several dorylaimid genera.

ENLARGEMENT

The progression from the anterior to the posterior pharyngeal section shows remarkable variation. It usually appears as a more or less gradual enlargement depending on the difference in diameter of the two sections (Fig. 6.3A-C). In particular taxa, however, both pharyngeal sections are either separated by a short, narrower, intermediate, isthmus-like section or abutting and with a deep constriction (Fig. 6.3D). Nair & Coomans (1973) richly illustrated these two conditions in members of the genus *Axonchium* (Fig. 6.3E).

BASAL EXPANSION AND PHARYNGEAL GLANDS

The posterior (basal) region of the pharynx shows great variability in its relative length. It very often occupies two-fifths to one-half of the total neck[3] length (Fig. 6.4A-C), but it may reach two-thirds to three-fourths of the neck length in representatives of a few genera (Fig. 4D). Some genera of a few families (*e.g.*, Leptonchidae, Longidoridae and Tylencholaimellidae), however, are characterised by having a remarkably short pharyngeal expansion, termed the pharyngeal bulb[4], whose length does not reach one-third (commonly one-fourth to one-fifth) of the total neck length (Fig. 6.4E). This pharyngeal bulb is usually cylindrical (Fig. 6.4F), but more occasionally it becomes pear-shaped (pyriform) (Fig. 6.4G). It may (Fig. 6.4G) or not (Fig. 6.4F) be separated from the anterior portion of the pharynx by a constriction and/or bear a small cavity inside, termed the triquetrous chamber, surrounded by a thicker inner lining (Fig. 6.1D).

[3] It is here assumed that the neck is the body region running from the anterior end to pharyngeal base. It is not identical to the pharynx, this strictly starting at the base of the stoma.

[4] The term 'pharyngeal bulb' has been traditionally used as synonymous with the pharyngeal expansion. Nevertheless, it is herein used in a more restricted sense to refer to those cases in which the pharyngeal expansion is very short.

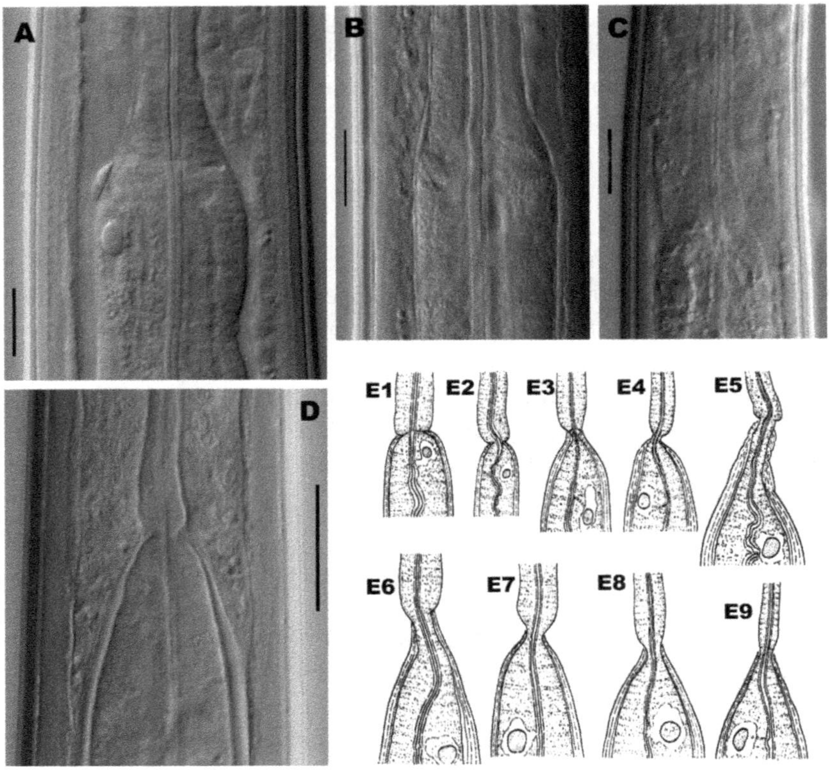

Fig. 6.3. *Pharyngeal enlargement. A-C: Gradual; D, E3, E4: Constriction; E1, E2: Constriction with abutting pharyngeal sections; E5-E9: Isthmus-like intermediate section. A:* Dorydorella bryophila *from the USA. B:* Labronema neopacificum *from Vietnam. C:* Longidorella deliblatica *from Spain. D:* Syncheilaxonchium nairi *from Iran. E:* Axonchium *spp. (After Nair & Coomans (1973),* Nematologica *18(1972), p. 501, Fig. 4A-I). (Scale bars: A, B, D = 20 μm; C = 10 μm.)*

The existence of a spiral muscular sheath (Fig. 6.5A-C) surrounding the pharyngeal basal expansion in members of the family Belondiridae (Hechler, 1969a) and particular genera of other families, for example, *Salimella* in Qudsianematidae or *Nygellus* in Nygellidae, deserves special mention and probably represents another good example of evolutionary convergence in dorylaims. As shown by Nair & Coomans

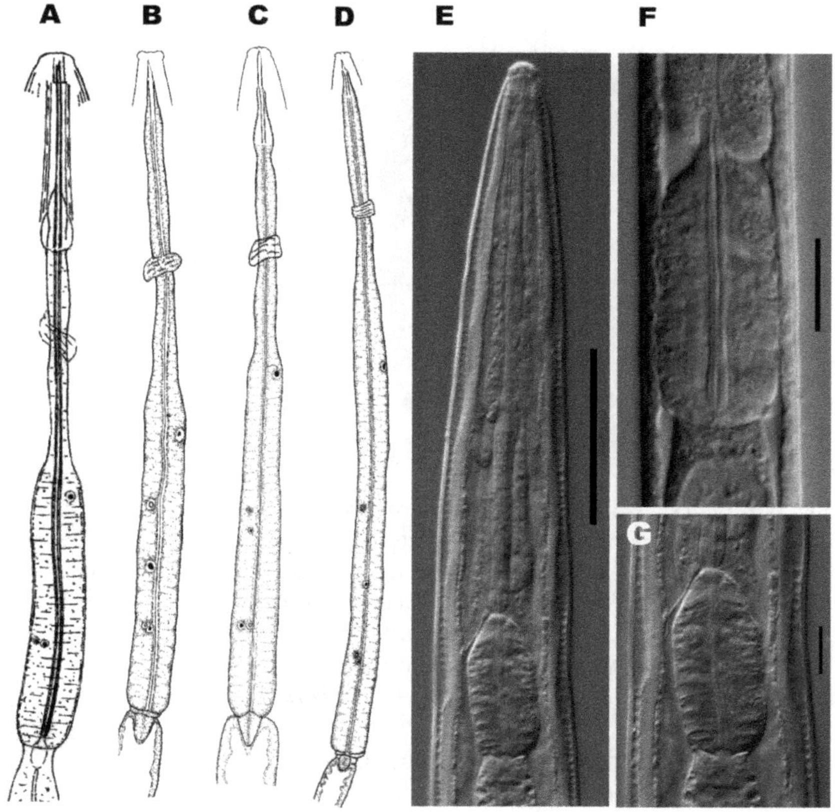

Fig. 6.4. *A-E: Pharynx of several dorylaimid taxa differing in relative length of their basal expansion; F, G: Pharyngeal bulb. A:* Enchodelus carpaticus *from Romania. B:* Aporcella charidemiensis *from Spain (after Álvarez-Ortega & Peña-Santiago (2016),* Nematology *18, p. 813, Fig. 1C). C:* Nevadanema nevadense *from Spain (after Álvarez-Ortega & Peña-Santiago (2012),* Nematology *12, p. 254, Fig. 3C). D:* Sectonema demani *(after Peña-Santiago & Álvarez-Ortega (2014),* Zootaxa *3881, p. 66, Fig. 2D). E, G:* Proleptonchus *sp. from Spain. F:* Oostenbrinkia *sp. from Iran. (Scale bars: E = 50 μm; F, G = 10 μm.)*

(1973) for *Axonchium* species, the bundles[5] forming the sheath are almost straight and parallel to the longitudinal axis of the pharynx

[5] The muscular bands of the spiral sheath are perceptible under LM only in well-preserved specimens, but the limits or margins of the sheath are readily distinguishable around the pharyngeal expansion.

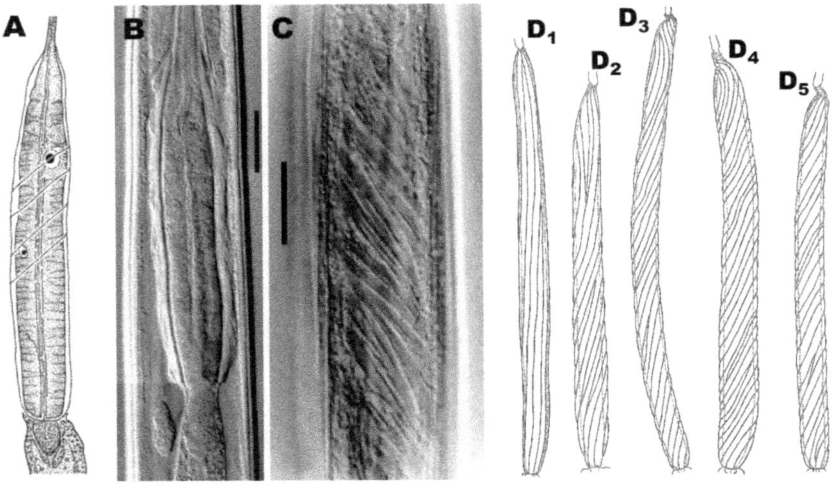

Fig. 6.5. *Pharyngeal basal expansion bearing spiral muscular sheath. A: Line diagram showing general aspect of spiral sheath; B: Strong sheath as observed in median view of* Lindseyus heterurus *from Spain (after Jiménez-Guirado & Murillo-Navarro (2004),* Journal of Nematode Morphology and Systematics *7, p. 93, Fig. 1C, D); C: Same in submedian view; D: Variation in arrangement of spiral muscular bands in* Axonchium *spp. (after Nair & Coomans (1973),* Nematologica *18(1972), p. 503, Fig. 5). (Scale bars: B, C = 20 μm.)*

(Fig. 6.5D_1) or slightly twisted, but they are spiral in others, being a diagnostic feature for species (see also Hechler, 1969b): dextrally spiralled in the posterior two-fifths, but almost straight anteriorly (Fig. 6.5D_2); dextrally spiralled in the posterior three-fourths, but almost straight or twisted sinistrally anteriorly (Fig. 6.5D_3); dextrally spiralled almost completely except at the anteriormost part where the bundles are sinistral (Fig. 6.5D_4); or dextrally spiralled over the entire length (Fig. 6.5D_5). Finally, some discolaims (Qudsianematidae, Discolaiminae) are characterised by having a thin, non-muscular sheath enveloping the pharyngeal expansion, often forming small but perceptible pockets at its base, which have occasionally been interpreted as a component of the pharyngo-intestinal junction (*cf.* Yeates, 1972).

As already mentioned, the expanded portion of the pharynx typically contains five unicellular glands (Fig. 6.1A) that are difficult to observe, but whose nuclei and nucleoli, as well as their outlets into the pharyngeal lumen, are easily seen in well-preserved specimens. One gland lies in the dorsal sector and is the anteriormost and the most conspicuous of the

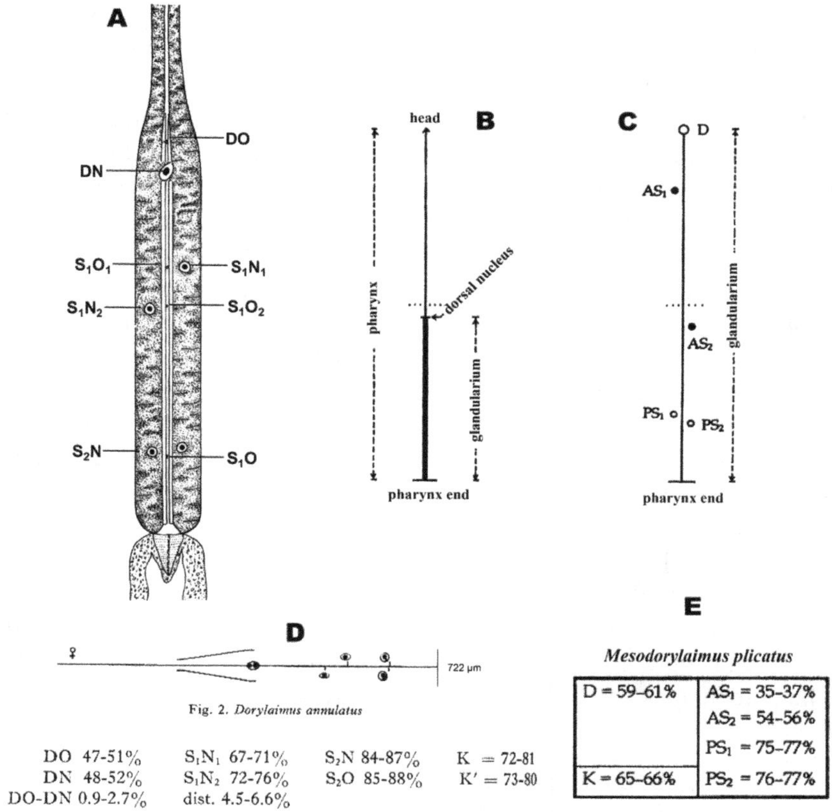

Fig. 6.6. *Pharyngeal gland nuclei and their location. A: Schematic (dorsal) view of dorylaimid pharyngeal expansion showing position of its gland nuclei and outlets (after Jairajpuri & Ahmad (1992),* Dorylaimida, *p. 27, Fig. 14C, modified); B, C: Andrássy's diagrams to represent the position of pharyngeal gland nuclei (after Andrássy (1998).* Opuscula Zoologica Budapestinensis *31, p. 166, Fig. 1A, B); D: Loof & Coomans' diagram to represent the position of pharyngeal gland nuclei and outlets and their corresponding percentages and ratios as calculated for* Dorylaimus annulatus *(after Loof & Coomans (1970),* Proceedings of the IX International Nematology Symposium *(*Warsaw, Poland, 1967*: p. 82, Fig. 2). E: Percentages representing position of pharyngeal gland nuclei of* Mesodorylaimus plicatus *following Andrássy's (1998) proposal.*

five, whereas the four other glands lie in the ventrosublateral sectors, two in each sector, forming two pairs, the posteriormost of which is located not far from the base of the pharynx (Fig. 6.6A). In some specialised taxa, the number of pharyngeal glands becomes reduced to three, the first (anterior) pair being absent. The existence and location of these glands was noted by several pioneer authors (*cf.* Jairajpuri & Ahmad, 1992), but it was Loof & Coomans (1970, 1972) who studied in depth their variability in dorylaims, proposed a scheme to represent (describe) their location and discussed their taxonomical interest. According to these authors, the dorsal gland[6] has a large and oblong nucleus with globular nucleolus, and its orifice always lies anterior to the nucleus, both being separated by a variable distance that is fairly constant within genera and species. The nuclei of the first pair are in general oblong-shaped too, have round nucleoli, do not lie at the same level (but are some distance apart), and their ducts run almost transversely, thus lying at the same level as their corresponding nucleoli. The nuclei of the second pair, always much larger than those of the first pair, are more or less globular, sometimes slightly transverse ovoid, with round nucleoli, occur at about the same level, and have their orifices located a short distance posterior to the nuclei. Loof & Coomans (1970, 1972) calculated the location of the pharyngeal gland nuclei and their orifices as percentages of the total neck length (*i.e.*, the distance from the anterior end to the pharyngeal base), and represented them by mean of simple diagrams (Fig. 6.6D). The percentages and ratios as originally proposed by Loof & Coomans are:

DO: orifice of dorsal gland;
DN: nucleus of dorsal gland (centre of nucleolus);
S_1O: orifice of ventrosublateral glands of the first pair;
S_1N: nuclei of these glands; the anterior one may be indicated as S_1N_1, the other as S_1N_2;
S_2O: orifice of the ventrosublateral glands of the second pair;
S_2N_2: nuclei of these glands;
K: distance DN-S_1N_1, expressed as percentage of the distance DN-S_1N_2; and
K': distance DN-S_2N_1, expressed as percentage of the distance DN-S_2N_2.

The proposals of Loof & Coomans on this matter have gained general acceptance, being much used by subsequent authors. After studying

[6] Robertson & Wyss (1979) elucidated in great detail the ultrastructure and function of the pharyngeal dorsal gland in *Xiphinema index*.

several dozen dorylaimid species, these authors (p. 159) concluded that the arrangement of the pharyngeal gland nuclei and their orifices were not only useful to characterise families, but also, to a certain extent, genera, and may be an important aid in identification. Nevertheless, the same authors did not find any correlation between such arrangement and the feeding habits of dorylaimid taxa. Andrássy (1998) presented a simpler representation scheme for the pharyngeal glands, with special emphasis on the location of their nuclei and no attention to their outlets (Fig. 6.6B, C). The abbreviations D, AS_1, AS_2, PS_1 and PS_2 were used instead of Loof & Coomans' values DN, S_1N_1, S_1N_2, S_1N, S_2N_2, respectively, and the term *glandularium* was introduced to name the region of pharyngeal expansion running from the dorsal nucleus to the pharyngeal base. With this new method, D is identical to DN, but the location of AS_1, AS_2, PS_1 and PS_2 are now expressed as percentages of *glandularium* length (Fig. 6.6E).

Basic patterns

Combining several variations in the nature of the anterior and posterior sections as well as their enlargement/separation, Jairajpuri & Ahmad (1992) distinguished five basic patterns of dorylaimid pharynx (Fig. 6.7) that correspond to the major taxonomical groups (superfamilies and suborders) as traditionally accepted.

In the dorylaimoid type (Fig. 6.7A), typical of the superfamily Dorylaimoidea, the two pharyngeal sections are muscular and comparable in length, the enlargement appearing more or less gradual, occasionally abrupt.

The presence of three large cardiac cells or glands at the pharyngo-intestinal junction (see below) distinguishes the nygolaimoid type (Fig. 6.7B), typical of the suborder Nygolaimina, from the dorylaimoid type. Belondirid (family Belondiridae) and other specialised genera are characterised by the presence of a spiral muscular sheath surrounding the pharyngeal expansion; the pharyngeal enlargement occurs either suddenly (abrupt expansion) or both sections are separated by a constriction or a short, isthmus-like portion – the belondiroid pattern (Fig. 6.7C). In many representatives of the superfamily Tylencholaimoidea (= Leptonchoidea), for example, members of the families Leptonchidae and Tylencholaimellidae, the anterior pharyngeal section becomes very slender and non-muscular, whereas the posterior one is rather short and often in

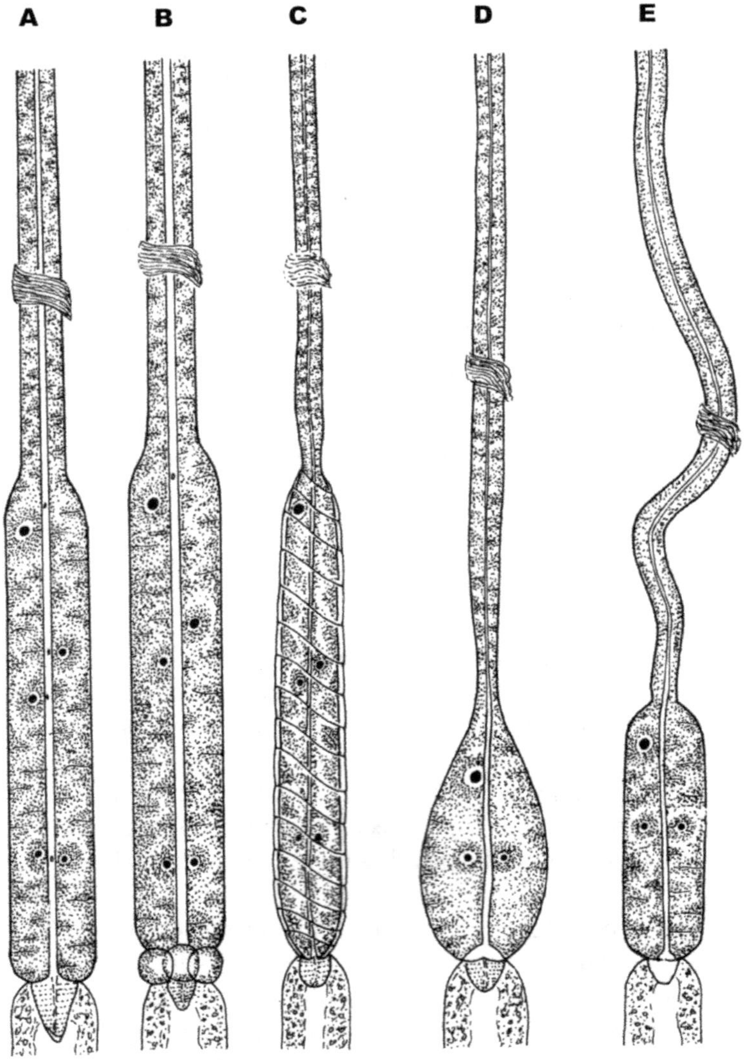

Fig. 6.7. *Pharyngeal patterns in dorylaims. A: Dorylaimoid; B: Nygolaimoid; C: Belondiroid; D: Leptonchoid; E: Longidorid. (After Jairajpuri & Ahmad (1992),* Dorylaimida, *p. 27, Fig. 14D-H.)*

the form of a pyriform bulb, both sections being connected by an abrupt enlargement or separated by a constriction; the number of pharyngeal glands in the bulb is reduced to three – the leptonchoid (tylencholaimoid) pattern (Fig. 6.7D).

Finally, the longidorid pattern, typical of the family Longidoridae (Fig. 6.7E), resembles the leptonchoid one but the basal bulb is cylindrical and much longer than wide.

However, this scheme of dorylaimid pharyngeal patterns should be interpreted with caution for two reasons. First, it is often possible to find many exceptions to the respective pattern in the representatives of the corresponding taxa. Second, these taxa are probably not natural (monophyletic) groups, the similarity observed in their pharynx being likely caused, at least in some cases, by the processes of evolutionary convergence.

Pharyngo-intestinal junction

The pharyngeal base connects to the anterior end of the intestine. Their junction is guarded by the cardia, a pharyngo-intestinal valve derived from the posteriormost part of the stomodeum and lined with cuticle (Nair & Coomans, 1973).

Its shape varies from subspherical (Fig. 6.8A_1), cardiform (Fig. 6.8A_2) to clavate (Fig. 6.8$A_{3,4}$) or long, cylindroid and tongue-like (Fig. 6.8A_{5-6}, B). Its function is to control the food flow from the pharynx to the intestine and to prevent regurgitation. The cardia is typically surrounded by intestinal tissue whose anterior end embraces it, sometimes forming a conical extension that bulges into the intestinal lumen (Fig. 6.8C).

This basic pattern has been modified in particular taxa, where special modifications occur. Thus, the members of the suborder Nygolaimina are characterised by the presence of three large (cardiac) cells or glands around the cardia at the junction between the pharyngeal base and the intestine (Fig. 6.8D, E). The existence of these cardiac cells is not exclusive to nygolaims as they have been also observed in many representatives of the genus *Oriverutus*, as well as in individual species of other taxa. Among nygolaims, the three cardiac glands are ovoid, almost always wider than long, but longer than wide in the genus *Paravulvus*. Occasionally, the intestine does not completely envelop the cardia but attaches to the latter a short distance posterior to the pharyngo-intestinal junction (Fig. 6.8F), only exceptionally attaching the cardia at its posterior end, as observed in the very rare genera *Qudsiella* and *Swangeria* (Fig. 6.8G).

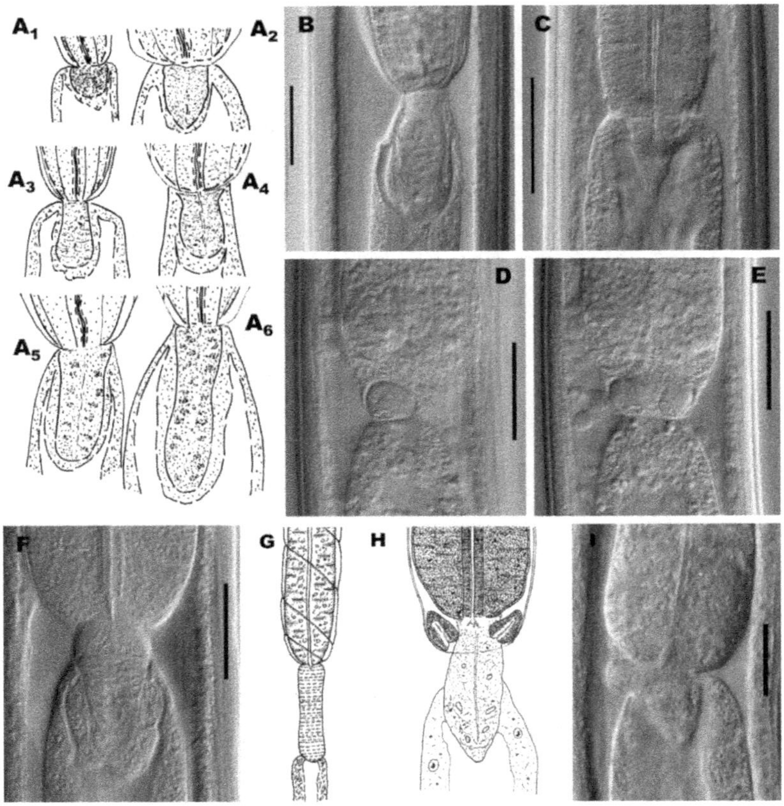

Fig. 6.8. *Pharyngo-intestinal junction. A: Variation in general aspect of cardia in* Axonchium *species: subspherical (A_1), cardiform (A_2) to clavate (Fig. $A_{3,4}$), and long and cylindroid, tongue-like ($A_{5,6}$) (after Nair & Coomans (1973),* Nematologica *18(1972), p. 500, Fig. 3F-K); B: Tongue-like cardia of* Axonchoides smokyensis *from the USA; C: Cardia of* Nevadanema nevadense *from Spain, with conical projection into the intestinal lumen; D, E: Cardiac glands/cells around pharyngo-intestinal junction in* Nygolaimus brachyuris *from Iran; F: Intestine partially embracing cardia in* Crassolabium persicum *from Iran; G: Intestine joining posterior end of cardia in* Swangeria longicauda *from Nicaragua (after Ahmad & Naz (2010),* Nematology *12, p. 887, Fig. 5F, modified); H: Strong belt-like structure surrounding pharyngo-intestinal junction in* Carcharolaimus eximius *from Spain (after Peña-Santiago & Liébanas (1994),* Fundamental and Applied Nematology *17, p. 407, Fig. 1G); I: Ring-like structure surrounding pharyngo-intestinal junction in* Oriverutus costaricensis *from Costa Rica, showing a dorsal lobe. (Scale bars: B, D-F = 20 μm; C = 50 μm; I = 10 μm.)*

The presence of a disc-like structure separating the pharyngeal base from the cardia has been repeatedly reported in the literature (for example, Heyns, 1965; Yeates, 1972; Jairajpuri & Ahmad, 1992), but this is probably a more or less developed ring-like structure, as noted by De Ley *et al.* (1993) in *Aporcelaimellus obtusicaudatus*. Such a ring-like structure may become strongly developed, forming a belt-like structure, as noted in species of the genus *Carcharolaimus* (Fig. 6.8H), and occasionally it becomes asymmetrical because its dorsal part is more developed, thereby forming a lobe (Fig. 6.8I).

In the only available monographic contribution on the matter, Yeates (1972) surveyed 58 dorylaimid species, made sections of eight of them, and proposed seven basic types of pharyngo-intestinal junction (Fig. 6.9), which differ in the nature of the valve, presence/absence of glands, disc (but see above), sheath, *etc.* (see also Jairajpuri & Ahmad, 1992). These types corresponded to several familiar taxa as conceived at that time: *i*) actinolaimid (valve moderately developed with narrow lumen, shield and sheath present); *ii*) aporcelaimid (valve moderately developed with narrow lumen and sheath sometimes present, disc present); *iii*) belondirid (valve large to massive with wide lumen and no sheath); *iv*) dorylaimid (valve well developed, with narrow lumen and bearing a terminal, non-muscular sheath); *v*) leptonchid (valve small with wide lumen and no terminal sheath); *vi*) longidorid (valve reduced or absent); and *vii*) nygolaimid (valve small with narrow lumen and no sheath, surrounded by three glands). Yeates' proposal, however, has not been very widely used to describe the nature of pharyngo-intestinal junction, perhaps due to the many exceptions found in the constituents of the corresponding families (*cf.* Jairajpuri & Ahmad, 1992).

Intestine proper

The intestine is a simple, tube-like structure made up of a single layer of epithelial cells. The cells are large and polygonal, each with a conspicuous nucleus and a series of cell inclusions. The intestine is separated from the body cavity by a basement membrane, whereas the plasma membrane is provided with numerous digitate projections termed *microvilli* (Jairajpuri & Ahmad, 1992). The number of intestinal cells may be few (oligocytous) or many (polycytous), the number varying depending on the taxonomical group considered. Contrary to other

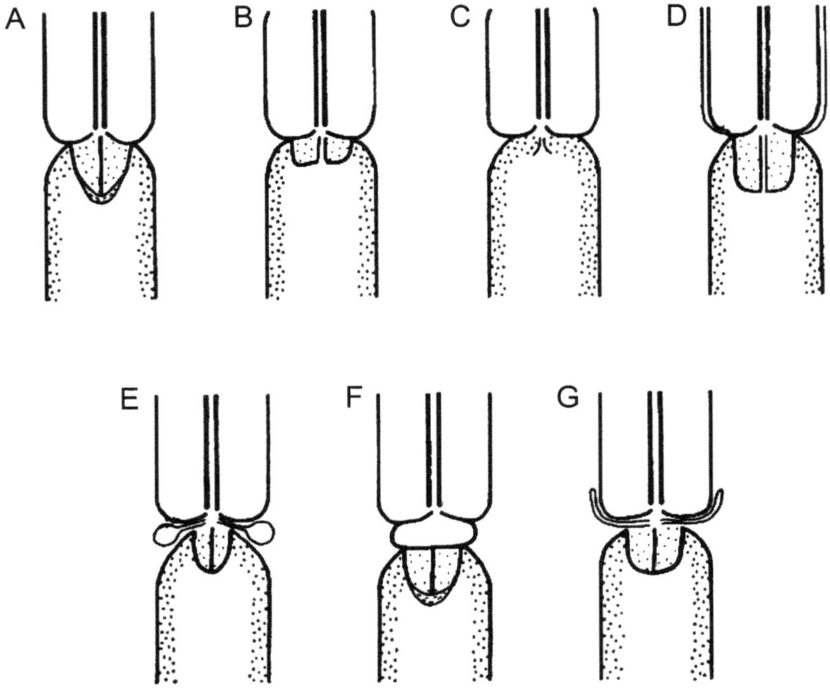

Fig. 6.9. *Types of pharyngo-intestinal junction. A: Dorylaimoid; B: Leptonchid; C: Longidorid; D: Belondirid; E: Nygolaimid; F: Aporcelaimid; G: Actinolaimid. (After Yeates (1972),* Journal of Natural History *6, p. 353, Fig. 8.)*

nematodes, the intestine of dorylaims is invariably divided into two regions: one anterior, the 'intestine proper' ('intestine' in the following section), and another posterior, the 'prerectum'. Such a division is a remarkable autapomorphic condition of the group. Both regions are often readily distinguishable, even under low LM magnification, because of differences in their aspect (colour) and contents of their cells, which are usually thinner and less dark in the prerectum.

The intestine is nearly always much longer than the prerectum, but a few exceptions do exist. It does not show special differentiations throughout its length. Very occasionally it bears a more or less developed, hollow and blind bulge or pouch (*caecum* or *diverticulum*) extending anteriad in parallel with the pharyngeal expansion, as observed in species of the genera *Vanderlindia* (Fig. 6.10A) and *Neometadorylaimus*.

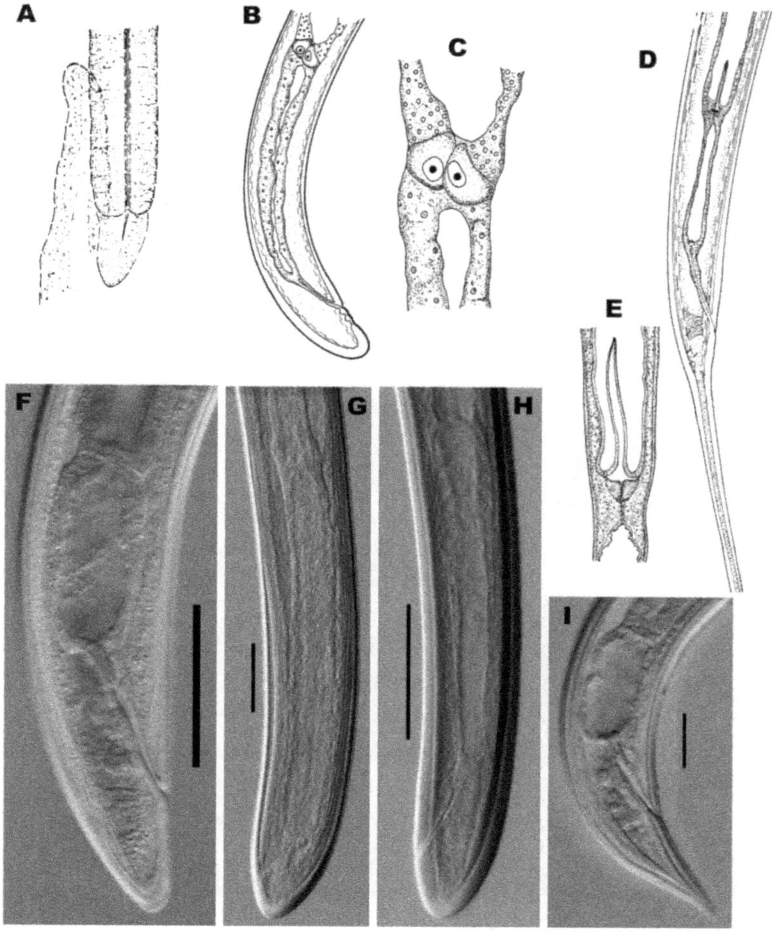

Fig. 6.10. *Intestine proper and prerectum. A: Anterior region of intestine of* Vanderlindia hispanica *from Spain, showing an anterior* caecum *(blind pouch) (after Jiménez-Guirado (1990),* Graellsia *46, p. 4, Fig. 1C); B, C: Guard cells at intestine-prerectum junction of* Dorylaimoides ornatus *from Spain (after Peralta & Peña-Santiago (1995),* Fundamental and Applied Nematology *18, p. 46, Fig. 5C, G); D, E: Intestine-prerectum junction of* Prodorylaimus longicaudatoides, *with an anteriorly projecting conical organ (after Abolafia* et al. *(1998),* Russian Journal of Nematology *6, p. 155, Fig. 3E, F); F-I: Variation of prerectum length in* Aporcelaimellus deserticola *from Spain (F),* Enchodelus carpaticus *from Romania (G),* Nevadanema nevadense *from Spain (H) and* Oriverutus costaricensis *from Costa Rica (I). (Scale bars: F, G = 50 μm; H = 100 μm; I = 20 μm.)*

The lumen of both intestine and prerectum are not totally continuous as their junction is demarcated by of a series of cells forming a transverse, septum-like structure, probably with a valvular function. The presence of three so called 'guard cells' (Fig. 6.10B), one dorsal and two subventral, has been repeatedly mentioned at the intestine-prerectum junction in representatives of several families. Also, a conical or tongue-like hollow projection, starting at the intestine-prerectum junction and anteriorly directed into the intestinal lumen (Fig. 6.10C), has been described in members of the family Dorylaimidae.

Prerectum

It is rather variable in length (Fig. 6.10D-G), but sometimes exhibits diagnostic value. Often, it appears comparatively short (up to three times the anal/cloacal body diameter long, Fig. 6.10D, G), but is occasionally longer (up to ten times the anal/cloacal body diameter, Fig. 6.10E, F) and, very exceptionally, as in a few members of the family Leptonchidae, it becomes larger (longer) than the intestine.

Prerectum length is commonly similar in both sexes but, in some cases, it may be appreciably longer in males than in females (Jairajpuri & Ahmad, 1992). Leaving aside its size (length), the prerectum does not show other particular specialisations or modifications. Very exceptionally, some species of a few genera, for example, *Discolaimus* and *Eudorylaimus*, have a variably developed blind sac on its posterior dorsal side that extends beyond the prerectum-rectum junction (Fig. 6.11A). The occurrence of unknown objects, other than microorganisms, in the prerectum of *Longidorus pisi* has been repeatedly reported (Jacobs & Heyns, 1982; Heyns *et al.*, 1984) but the precise function of the prerectum remains unknown.

Rectum

The hindermost part of the alimentary canal is the rectum, which displays a marked sexual dimorphism. In females it is a simple tube running from its junction to the prerectum until the anus, terminating in an opening on the body surface (Fig. 6.11A-C). The prerectum-rectum junction is marked by a conspicuous narrowing, which probably serves as a valve, and appears surrounded by a muscular ring (sphincter).

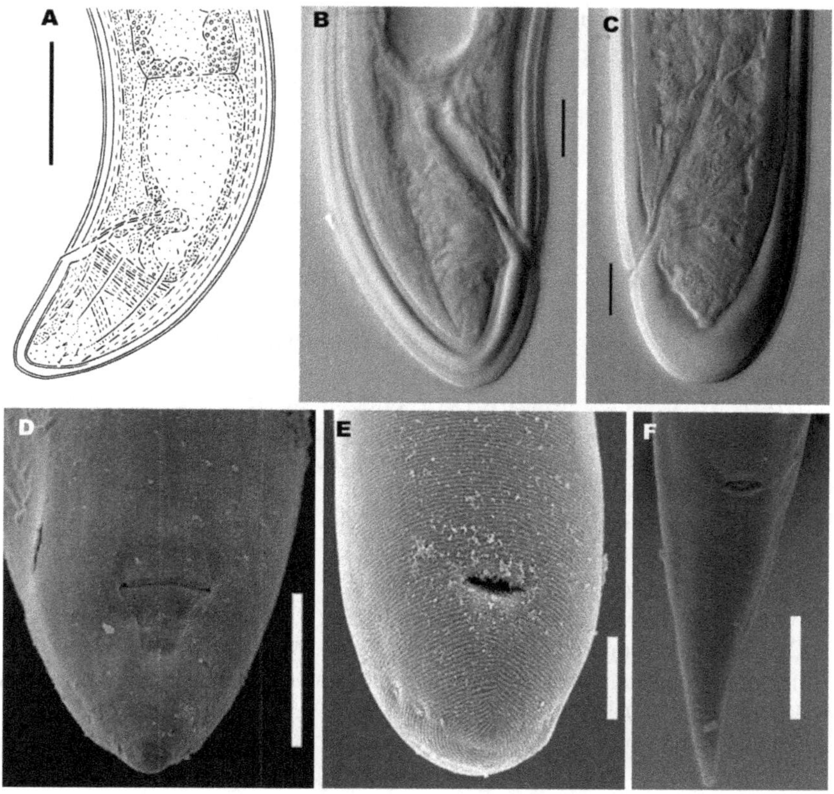

Fig. 6.11. *Prerectum-rectum junction, rectum and anus. A: Prerectum of* Discolaimus paramajor *(after Coomans (1966).* Revue de Zoologie et Botanie Africaines *74, p. 292, Fig. 3E), with a posterior blind sac; B: Rectum of* Crassolabium vietnamense *from Vietnam; C: Rectum of* Syncheilaxonchium nairi *from Iran; D-F: Ventral view of posterior body region of* Labronema angeloi *from Spain (D),* Labronemella labiata *from Iran (E) and* Oriverutus asaccatus *from Costa Rica (F), showing anus. (Scale bars: A = 30 μm; B, C, E, F = 10 μm; D = 20 μm.)*

Rectum length is usually about equal to body diameter at this level, occasionally longer, the latter situation then having taxonomical or diagnostic value.

The rectum is derived from the ectoderm during embryonic development and is internally lined with cuticle (Jairajpuri & Ahmad, 1992). In males, the rectum is very short and joins the ejaculatory duct resulting in a common duct, the cloaca, which contains the copulatory apparatus and opens to the body surface *via* the cloacal aperture. Both the anus and the cloacal aperture are located in a ventral position and show similar morphology, appearing as a transverse, straight or slightly arched, narrow or somewhat oval slit (Fig. 6.11D-F).

References

ABOLAFIA, J., PEÑA-SANTIAGO, R. & JIMÉNEZ-GUIRADO, D. (1998). Nematodes of the order Dorylaimida from Andalucía Oriental, Spain. The genus *Prodorylaimus* Andrássy, 1959. II. Bisexual species. *Russian Journal of Nematology* 6, 147-159.

AHMAD, W. & NAZ, T. (2010). Description of new and known species of the rare nematode genera *Qudsiella* Jairajpuri, 1967 and *Swangeria* Thorne, 1939 (Belondiridae: Dorylaimida). *Nematology* 12, 877-890. DOI: 10.1163/138855410X494152

ALI, S.M. & PRABHA, M.J. (1971). A case of an abnormal oesophagus in a juvenile specimen of *Thornenema baldum*. *Nematologica* 16(1970), 601-602. DOI: 10.1163/187529270X00838

ÁLVAREZ-ORTEGA, S. & PEÑA-SANTIAGO, R. (2012). Nematodes of the order Dorylaimida from Andalucía Oriental, Spain. *Nevadanema nevadense* gen. n., sp. n. (Qudsianematidae) from Sierra Nevada National Park. *Nematology* 14, 249-264. DOI: 10.1163/156854111-00002995

ÁLVAREZ-ORTEGA, S. & PEÑA-SANTIAGO, R. (2016). *Aporcella charidemiensis* sp. n. (Dorylaimida: Aporcelaimidae) from the southern Iberian Peninsula, with comments on the phylogeny of the genus. *Nematology* 18, 811-821. DOI: 10.1163/15685411-00002995

ANDRÁSSY, I. (1998). Once more: the oesophageal gland nuclei in the dorylaimoid nematodes. *Opuscula Zoologica Budapestinensis* 31, 165-170.

COOMANS, A. (1966). Some nematodes from Congo. *Revue de Zoologie et Botanie Africaines* 74, 287-312.

COOMANS, A., DE CONINCK, L. & HEIP, C. (1978). Data to be considered in descriptions of new species or redescription of poorly known species. *Annales de la Société Royal de Zoologie de Belgique* 108, 119-122.

DE LEY, P., LOOF, P.A.A. & COOMANS, A. (1993). Terrestrial nematodes from the Galápagos Archipelago II: Redescription of *Aporcelaimellus obtusicaudatus* (Bastian, 1865) Altherr, 1968, with review of similar species and a nomenclature for the vagina in Dorylaimida (Nematoda). *Bulletin de l'Institut Royal des Sciences Naturelles de Belgique* 63, 13-34.

GROOTAERT, P. & COOMANS, A. (1981). The formation of the anterior feeding apparatus in dorylaims. *Nematologica* 26(1980), 406-431. DOI: 10.1163/187529280X00350

GROOTAERT, P. & WYSS, U. (1978). Ultrastructure and function of the anterior feeding apparatus in *Labronema* sp. *Nematologica* 24, 243-250. DOI: 10.1163/187529278X00209

HECHLER, H.C. (1969a). Variability of the basal esophageal sheath in belondirid nematodes. *Journal of Nematology* 1, 160-165.

HECHLER, H.C. (1969b). Taxonomy and morphology of *Axonchium* (Nematoda: Belondiroidea), and a description of *A. thornei* n. sp. *Journal of Nematology* 1, 321-348.

HEYNS, J. (1965). On the morphology and taxonomy of the Aporcelaimidae, a new family of dorylaimid nematodes. *Entomology Memoirs, Department of Agriculture Technical Services, Republic of South Africa* 10, 1-51.

HEYNS, J., JACOBS, P.J.F., LOOTS, G.C. & TIEDT, L. (1984). On the occurrence of unknown objects in the prerectum of *Longidorus pisi* Edward, Misra & Singh, 1964 (Nematoda: Longidoridae). *Phytophylactica* 16, 167-169.

JACOBS, P.J.F. & HEYNS, J. (1982). *Longidorus* species from sugar cane in Natal (Nematoda: Longidoridae). *Phytophylactica* 14, 195-204.

JAIRAJPURI, M.S. & AHMAD, W. (1992). *Dorylaimida. Free-living, predaceous and plant-parasitic nematodes*. New Delhi, India, Oxford & IBH Publishing Co. Pvt. Ltd.

JIMÉNEZ-GUIRADO, D. (1990). Descripción de *Vanderlindia hispanica* sp. n. (Nematoda: Tylencholaimidae). *Graellsia* 46, 3-5.

JIMÉNEZ-GUIRADO, D. & MURILLO-NAVARRO, R. (2004). Observations on *Lindseyus heterurus* (Schuurmans-Stekhoven & Teunissen, 1938) Coomans & Kheiri, 1986 (Nematoda: Belondiroidea) from Spain. *Journal of Nematode Morphology and Systematics* 7, 91-96.

LIPPENS, P.L. (1975). Ultrastructure of Dorylaimidae. In: Lamberti, F., Taylor, C.E. & Seinhorst, J.W. (Eds). *Nematode vectors of plant viruses*. London, UK, Plenum Press, pp. 159-178.

LIPPENS, P.L., COOMANS, A., DE GRISSE, A.T. & LAGASSE, A. (1974). Ultrastructure of the anterior body region in *Aporcelaimellus obtusicaudatus* and *A. obscurus*. *Nematologica* 20, 242-256. DOI: 10.1163/187529274X00212

LOOF, P.A.A. & COOMANS, A. (1970). On the development and location of the oesophageal gland nuclei in Dorylaimina. *Proceedings of the IX International Nematology Symposium* (*Warsaw, Poland, 1967*), pp. 79-161.

LOOF, P.A.A. & COOMANS, A. (1972). The oesophageal gland nuclei of Longidoridae (Dorylaimida). *Nematologica* 18, 213-233. DOI: 10.1163/187529272X00458

MUSHTAQ, P., BANIYAMUDDIN, M. & AHMAD, W. (2006). Studies on nematodes of the family Aulolaimoididae (Nematoda: Dorylaimida) with descriptions of five new species. *Nematology* 8, 761-780. DOI: 10.1163/156854106778877965

NAIR, P. & COOMANS, A. (1973). The genus *Axonchium* (Nematoda. Belondiridae). I. Morphology, juvenile stages, diagnosis and list of species. *Nematologica* 18 (1972), 494-513.

PEÑA-SANTIAGO, R. & ÁLVAREZ-ORTEGA, S. (2014). Re-description of three new species of the genus *Sectonema* Thorne, 1930 (Nematoda; Dorylaimida: Aporcelaimidae) originally studied by E. Altherr. *Zootaxa* 3881, 63-74. DOI: 10.11646/zootaxa.3881.1.5

PEÑA-SANTIAGO, R. & LIÉBANAS, G. (1994). Nematodes of the order Dorylaimida from Andalucía Oriental, Spain. The genus *Carcharolaimus* Thorne, 1939 with description of *C. eximius* sp. n. and a compendium of the genus. *Fundamental and Applied Nematology* 17, 405-413.

PERALTA, M. & PEÑA-SANTIAGO, R. (1995). Nematodes of the order Dorylaimida from Andalucía Oriental, Spain. The genus *Dorylaimoides* Thorne & Swanger, 1936. 1. Didelphic species. *Fundamental and Applied Nematology* 18, 35-53.

ROBERTSON, W.M. & WYSS, U. (1979). Observations on the ultrastructure and function of the dorsal oesophageal gland cell in *Xiphinema index*. *Nematologica* 25, 391-396. DOI: 10.1163/187529279X00550

SIDDIQI, M.R. (1971). *Oriverutus lobatus* gen. n., sp. n. and *Sicaguttur sartum* gen. n., sp. n. (Nematoda: Dorylaimoidea) from cultivated soils in Africa. *Nematologica* 16(1970), 483-491. DOI: 10.1163/187529270X00667

TAYLOR, C.E., THOMAS, P.R., ROBERTSON, W.M. & ROBERTS, I.M. (1970). An electron microscope study of the oesophageal region of *Longidorus elongatus*. *Nematologica* 16, 6-12. DOI: 10.1163/187529270X00405

YEATES, G.W. (1972). The oesophago-intestinal junction in the Dorylaimoidea (Nematoda). *Journal of Natural History* 6, 343-355.

Nematology Monographs & Perspectives, 2021, Vol. 13, 81-105

Chapter 7

Female genital system

General concept and terminology

The female genital system of dorylaims typically comprises two functional ovaries (di-ovarian condition) with their corresponding genital tracts (didelphic condition), forming two genital branches that merge and lead to a common vagina opening on the body surface *via* the vulva. One of the two genital branches occupies an anterior position, *i.e.*, it runs from the vagina in an anterior direction, whereas the other is posterior, *i.e.*, it extends in a posterior direction from the vagina. One of the two ovaries may be lost or became vestigial (reduced) and non-functional (mono-ovarian condition). When only the anterior ovary is present and functional, such a condition is termed mono-pro-ovarian. Conversely, if only the posterior ovary is developed and functional, this is known as the mono-opistho-ovarian condition. Whenever one ovary is functional, its corresponding genital tract is always well developed and functional. In some cases, the genital tract corresponding to a lost or non-functional (vestigial) ovary is totally lacking, which means that only one ovary (mono-ovarian condition) and one genital tract (mono-delphic condition) exist. In other cases, the genital tract corresponding to a lost or non-functional ovary persists in adult females, occasionally retaining some of its functions; thus only one ovary exists (mono-ovarian condition) with two genital tracts (didelphic condition). Figure 7.1A illustrates a system consisting of two functional ovaries (di-ovarian) with (always) their corresponding genital tracts (actually didelphic, but it should be unnecessary to reiterate this description). Figure 7.1B shows a system with only one functional ovary (mono-ovarian), the posterior one (= mono-opistho-ovarian) and, in addition to its corresponding posterior genital tract, the anterior tract is also present and well developed (constituted by the uterus, sphincter and oviduct, see below), the system hence being referred to as mono-opistho-ovarian-didelphic. In Figure 7.1C only the posterior ovary is

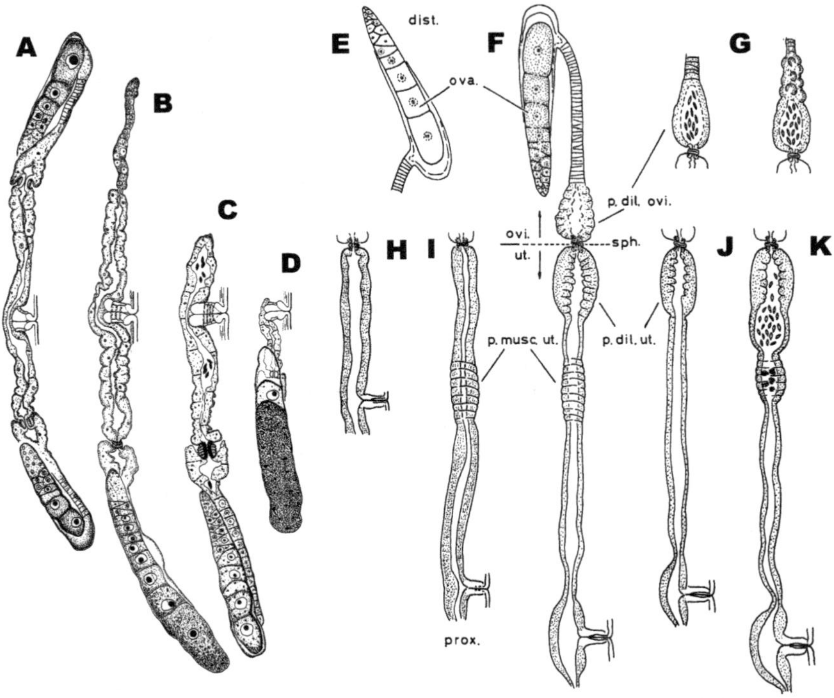

Fig. 7.1. *General morphology of female genital system. A-D: General types. A: Di-ovarian; B, C: Mono-opistho-ovarian-didelphic; D: Mono-opistho-ovarian-monodelphic. E-K: General scheme of the different components of the genital branch (after Coomans (1965),* Nematologica *10(1964), p. 619, Fig. 12). E: Ovary; F: Genital branch, entire; G: Enlarged part of oviduct functioning as spermatheca; H-K: Different types of uterus. (Abbreviations: dist. = distal, ovi. = oviduct, prox. = proximal, p. dil. ovi. =* pars dilatata oviductus, *p. dil. ut. =* pars dilatata uteri, *p. musc. ut. =* pars muscularis uteri, *sph. = sphincter, ut. = uterus.)*

functional although the anterior genital tract exists (even though visibly reduced), the system thus being mono-opistho-ovarian-didelphic. Figure 7.1D shows a mono-opistho-ovarian system with no trace of the anterior genital tract (actually mono-opistho-delphic, although a superfluous reiteration).

Each female genital branch typically consists of an ovary (Fig. 7.1E) and genital tract. The latter is a tube-like structure, often divided into two main sections (oviduct and uterus) separated by a narrowing encircled by a muscular ring (sphincter) (Fig. 7.1F). The oviduct, and especially the uterus, may bear other specialisations (Fig. 7.1G-K). Both uteri merge and join the vagina, which opens on the body surface *via* the vulva. Coomans (1965) provided an excellent study of the morphology of the female genital system and proposed a terminology for its components that is still valid. Thus, Coomans' ideas, along with some additions (Grimaldi de Zio *et al.*, 1975; Bleve-Zacheo *et al.*, 1976; Geraert *et al.*, 1980; Van de Velde & Coomans, 1988; Kruger & Heyns, 1990; Jairajpuri & Ahmad, 1992; De Ley *et al.*, 1993), form the basis of the following description.

The ultrastructure of the genital system was studied and analysed in detail, especially in longidorid forms, by Van de Velde *et al.* (1990a, b) and Coomans & Claeys (1998), who provided much very interesting information on this matter.

Ovary

The ovary consists of ovarian epithelial cells and germ-cells, the former existing as a thin layer around the latter (Figs 7.1E; 7.2A-C). Three ovarian regions are distinguishable. The germinative or multiplication zone occurs at the tip (*i.e.*, the distal end), where the germ-cells are arranged in several rows with many nuclei, yet lack perceptible cell membranes, and where the oocytes are formed. The growth zone, where the oocytes increase in size, is first arranged in a double and then in a single row. The ripening zone is at the proximal end of ovary and is where the oocytes accumulate yolk before passing into the oviduct.

The ovary is enveloped in a thin-walled sac, which is more distinct in its proximal region. The epithelial cells of the ovarian sac are easily seen only in the proximal half of the ovary, where an outermost layer of connective (?) tissue can be observed thanks to its superficial, circular plications. These plications, as well as the boundaries of the epithelial cells, disappear when the region becomes extended by the growing egg cells. In the distal region of the ovary the wall of the sac is very thin and difficult to observe. The ovary is always reflexed, *i.e.*, its orientation is such that the germinative zone is situated closer to the vulva than the growth zone. Due to its subterminal junction with the oviduct, there is a blind sac at the proximal end of the ovary where ripening egg cells remain until nearly full-grown.

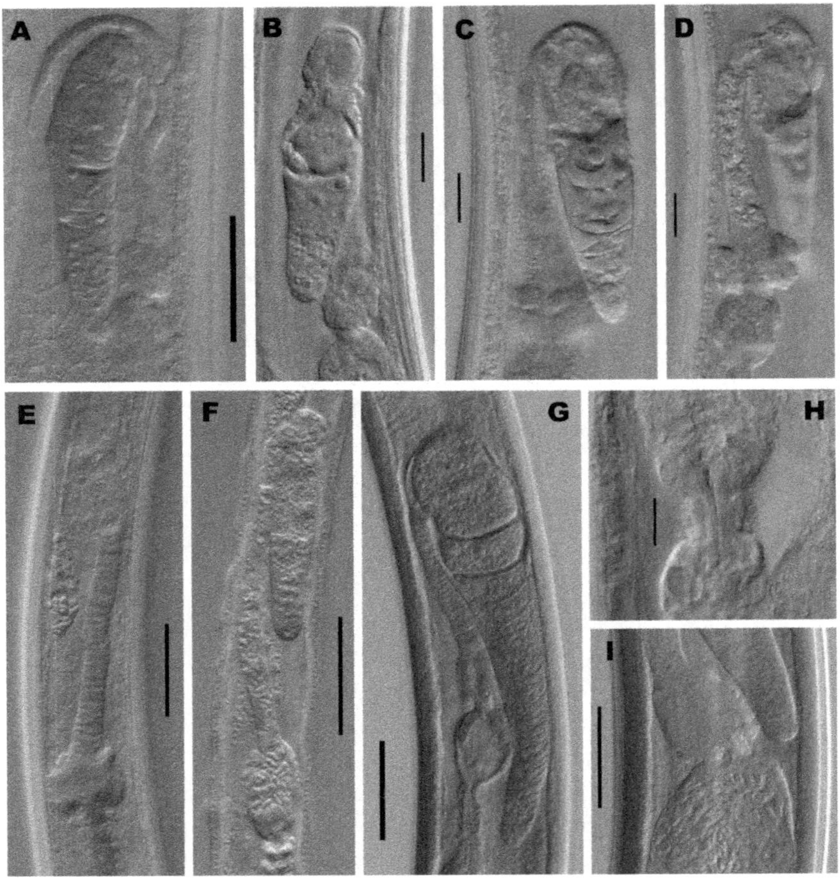

Fig. 7.2. *Female genital system. A-C: Ovary; D-G: Oviduct; H, I: Oviduct-uterus junction. A:* Aporcelaimellus baeticus *from Spain; B:* Oriverutus tropicus *from Costa Rica; C, D:* Oriverutus costaricensis *from Costa Rica; E:* Heterodorus geraldi *from Romania; F:* Proleptonchus *sp. from Spain; G:* Aporcelinus mamilatus *from Sri Lanka; H:* Nevadanema nevadense *from Spain; I:* Sectonema *sp. from Vietnam. (Scale bars: A-D, I = 20 μm; E, F = 50 μm; G = 40 μm; H = 10 μm.)*

The size (length) of the ovary depends on the developmental stage of the female and has not much taxonomic interest. It is often relatively small, shorter than the oviduct and does not usually reach the level of the sphincter (Fig. 7.2B, F), although it may do so (Fig. 7.2C, I) or even surpass this level (Fig. 7.2A, G).

Oviduct

The oviduct is the first section of the genital tract and joins the ovary subterminally. It consists of a narrow, distal part and an enlarged proximal one (Figs 7.1F, G; 7.2A, D-G). The distal part is always longer than the proximal one and occupies two-thirds to three-fourths of its total length. It is made of prismatic or relatively tall cells arranged in a longitudinal series (Fig. 7.2E), these cells not always being clearly visible and occasionally containing distinct granulation. A lumen is never conspicuous in this part of the oviduct. The enlarged, proximal part, termed the *pars dilatata oviductus*, is made of an epithelium of relatively shorter cells with obscure boundaries, and a distinct lumen is very often visible. The oviduct enlargement may be more or less gradual (Fig. 7.2G) or abrupt (Fig. 7.2E), although it may depend on the functional state of the female, whether fertilised or not, or bearing a ripened egg(s), *etc*.

Geraert *et al.* (1980), who provided a detailed description of the oviduct structure mainly based on TEM studies, found that both parts of the oviduct differ significantly in their histology. The distal part (oviduct *sensu stricto* after Geraert *et al.*, 1980) consists of a long column of consecutive disc-like cells, each with a relatively large nucleus. The total number of cells ranges between 17 and 36 in the species examined by the authors. Three or four ovarian cells surround the oviduct at its distal end where it enters the ovary, whereas its opposite end is surrounded by the proximal part, more or less bulging into the latter. The proximal part (the *pars dilatata* after Geraert *et al.* (1980) does not belong to the oviduct according to these authors) is sac-like and consists of two or three rings of spindle-shaped to elongated cells. The total number of cells ranges between 12 and 22, the ring around the oviduct being composed of 6-8 cells and that surrounding the sphincter of 6-12 cells.

The functional morphology of the oviduct was studied in depth by Coomans (1965) and Kruger & Heyns (1988). It drives ripened eggs from the ovary to the uterus, but the role in this process of each of the two oviduct parts is significantly different. Whenever an egg is present in the oviduct, it invariably occurs in the proximal, dilated part, never in the distal, narrow part. This means that the passage of the egg through the distal part is fast, a process that may be facilitated by secretions of the epithelial cells. The proximal, dilated part of the oviduct has a very extendible wall and is able to envelop the egg completely, serving as a 'resting chamber' where the last stages of the ripening process take place, fertilisation eventually happening here. It is also possible that this part plays some role in the formation of the egg shell. In bisexual species, fertilised females may bear or contain sperm cells throughout the oviduct (Fig. 7.2F), but more especially within the proximal dilated part that hence functions as a spermatheca (Fig. 7.1G).

Sphincter

The junction between the oviduct and the uterus is marked by a narrowing or constriction, almost always easily seen (Fig. 7.2H, I). Here the lumen of the genital tract becomes very reduced, whereas the wall lining is visibly plicate and sclerotised. The narrowing or constriction is surrounded by a circular muscle whose development may differ considerably depending on the species considered (compare Fig. 7.2H and Fig. 7.2I). Unfortunately, the nature and variability of this structure has not been studied in sufficient depth to establish recognisable patterns of variation.

Geraert *et al.* (1980) found that the sphincter consists of two parts. One lies between the uterus and the *pars dilatata* (*oviductus*), but may be surrounded by tissue from both. It has a sclerotised lumen lining, probably due to a strongly folded lumen wall, which in sections appears triangular and, occasionally, has one cell on each side. The second part is a ring of cells with large vacuoles, its lumen bordered either by the vacuoles or by cytoplasm and nuclei. The number of cells ranges between four and six.

Uterus

The morphology of the uterus is remarkably variable among dorylaimid taxa. It is very constant, however, within specimens of the same species, often having taxonomic interest for the separation of rather similar species.

According to Geraert *et al.* (1980), it is a hollow cylinder without recognisable cell boundaries, either appearing as a quadricolumella or as a tube with surrounding cells, the outer walls of which bulge. In addition, the uterine wall has well-developed circular and oblique muscular fibres

(Jairajpuri & Ahmad, 1992), and its inner lining is made of squamous epithelium that is variously modified in the different sections.

Basic types

In its simplest form, the uterus is a tube-like structure without any special differentiation (uni- or monopartite type, Figs 7.1H; 7.3A). Coomans (1965) suggested that this kind of uterus should be regarded as a primitive condition, although such a condition is not necessarily an indication of the primitive character of the species as paedogenetic phenomena may have occurred. In some cases, the uterine distal portion, close to the sphincter, becomes appreciably dilated (*pars dilatata uteri*), nearly spherical and with a wide lumen; the uterus therefore consists of two distinguishable sections or parts (bipartite type, Fig. 7.1J). In bisexual species the dilated part of the uterus often contains a mass of sperm cells, thereby functioning as a spermatheca. The uterine distal part, especially if it becomes dilated, certainly plays an important role in the formation of the egg shell together with the proximal part of the oviduct. In exceptional cases, for example, the members of the genera *Heterodorus* and *Oriverutus*, the distal part of the uterus is not distinctly dilated, but its inner lining appears especially differentiated and bears some refractive inclusions (granules, Fig. 7.3B), it being possible to distinguish the two uterine sections (*i.e.*, also a bipartite type).

Many species display a more complex uterus in which at least three parts or sections are recognisable (tripartite type, Fig. 7.1F, I, K). An intermediate (median), often short section, somewhat enlarged and provided with distinct circular muscles, is sometimes present (Figs 7.1I; 7.3C), the uterus thus comprising two broader (one distal and one proximal) sections separated by the median muscularised part (*pars musculosa uteri*) with a thicker wall and narrower lumen. In some cases, the *pars musculosa* is not in a median position but close to the *pars dilatata* (see below). Conversely, other species, for example, the members of the genus *Enchodelus* and selected representatives of many other genera, have tripartite uteri in which two (one distal and another proximal) smaller and dilated parts are separated by a longer and narrower intermediate section (Fig. 7.3D).

Occasionally, the combination of both differentiations (*pars dilatata* and *pars musculosa*) results in a more complex uterus, actually with four distinguishable sections (tetrapartite type, Fig. 7.1F). The proximal

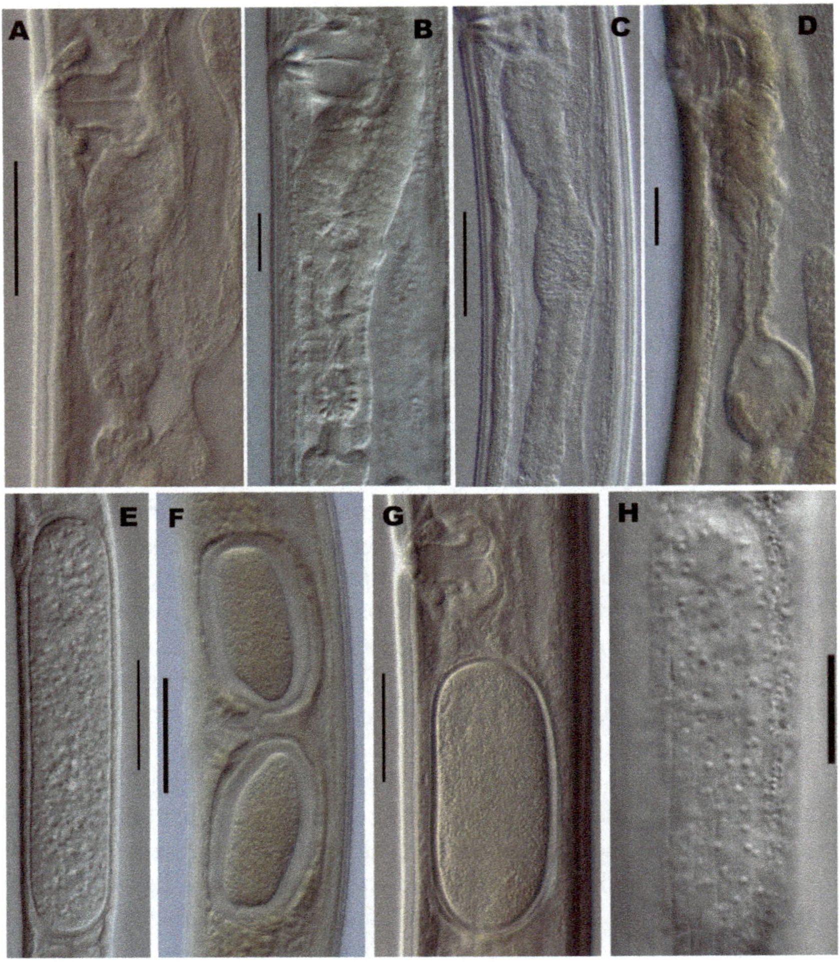

Fig. 7.3. *Female genital system: uterus and uterine egg. A: Simple (monopartite) uterus of* Nevadanema nevadense *from Spain; B: Bipartite uterus of* Oriverutus ecae *from Costa Rica; C: Tripartite uterus of* Labronemella labiata *from Iran; D: Tripartite uterus of* Aporcelinus mamillatus *from Sri Lanka; E: Uterine egg of* Oostenbrinkia *sp. from Iran; F: Uterine eggs of* Inbionema biforme *from Costa Rica; G: Uterine egg of* Nevadanema nevadense *from Spain; H: Uterine egg of* Metaporcelaimus ovogranulosus *from the USA. (Scale bars: A, C, F, G = 50 μm; B = 10 μm; D, E, H = 20 μm.)*

part of the uterus often shows a larger lumen and its wall lacks distinct musculature except near its junction with the vagina, where obliquely oriented muscles occur and converge on the vagina. The development of these muscles varies greatly from species to species and is often so weak that no striation is visible. Conversely, this region can become somewhat distinct from the rest of the uterus and forms, together with the corresponding region of the other uterus, a functional entity, the ovejector, which serves to control the movement of eggs during oviposition.

Uterine special differentiations

The uterus may also display other special differentiations. The presence of refractive inclusions (apophyses, spines, crystalloids, *etc.*) attached to the uterine wall is frequent in members of the genera *Metaxonchium* and *Labronema*, and reaches an extraordinary development and variability in the genus *Xiphinema*. Andrássy (1996) used the term 'echinophore uterus' to designate such a condition in *Metaxonchium* species. When present, the *pars musculosa* may become especially developed, very often bearing very distinct inclusions – the so-called 'Z-differentiation'[1] (Fig. 7.1K). The uterine differentiation has been exhaustively studied in the genus *Xiphinema*, where it displays remarkable interspecific but limited intraspecific variation (Luc, 1961; Grimaldi de Zio *et al.*, 1979; Kruger, 1988; Coomans *et al.*, 1992, 2001) (Fig. 7.4). Kruger (1988) found that, at that time, 48 *Xiphinema* species only possessed Z-differentiation, 14 species had both Z-differentiation and uterine spines, and nine species only had spine-shaped structures. The same author concluded that uterine inclusions can be divided into three basic types:

i) globular structures, often containing vesicles and apparently attached to the uterine wall by stem-like elements;
ii) uterine spines (spiniform structures), often associated with a Z-differentiation, more or less regularly distributed, but lacking in the

[1] Kruger (1988), who carried out an excellent revision of uterine differentiation in members of the genus *Xiphinema*, recommended the use of the expression 'Z-differentiation' instead of 'Z-organ' and 'pseudo-Z-organ' because of the difficulty in distinguishing them, they often being misinterpreted or misunderstood by many authors (*cf.* Luc, 1973, 1975).

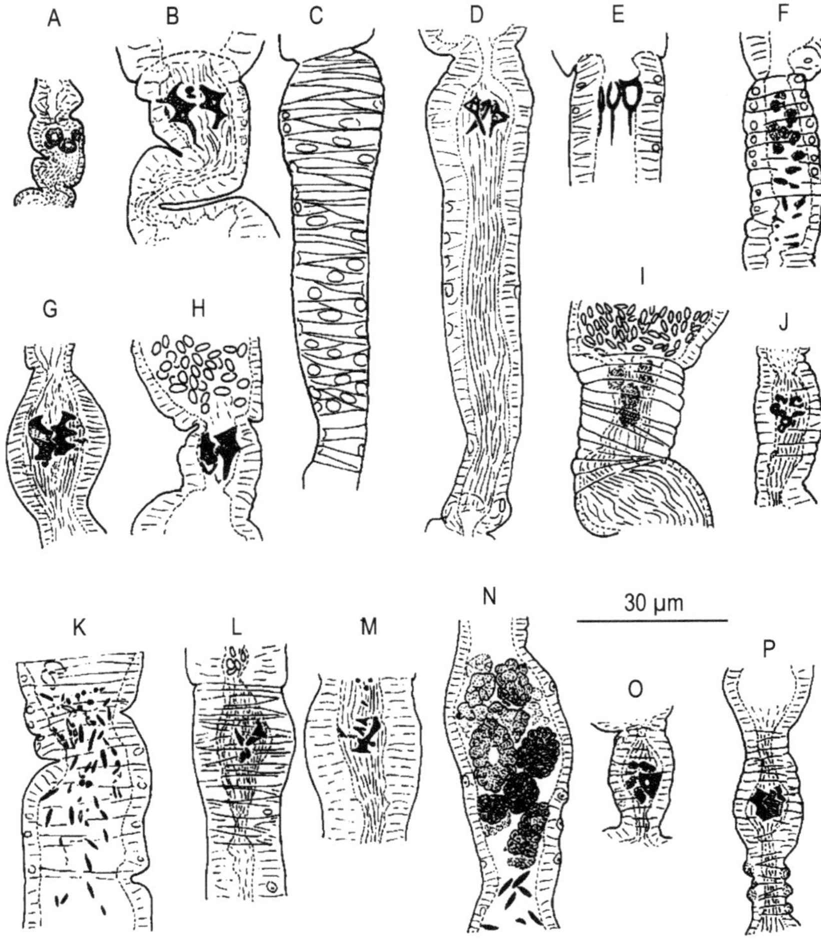

Fig. 7.4. *Female genital system: variability in uterine Z-differentiation in members of the genus* Xiphinema. *A:* X. stenocephalum*; B:* X. oryzae*; C-E:* X. hallei*; F:* X. malagasi*; G:* X. ifacolum*; H:* X. algeriense*; I:* X. limpopoense*; J:* X. limbeense*; K:* X. israeliae*; L, M:* X. imambaksi*; N:* X. macroacanthum*; O:* X. ebriense*; P:* X. manubriatum. *(After Coomans* et al. *(2001),* Annales Sciences Zoologiques, Musée Royal de l'Afrique Centrale, Tervuren, Belgium *287, p. 26, Fig. 10.)*

pars dilatata, sometimes concentrated adjacent to Z-differentiation and with their spiniform processes consistently oriented towards the *pars dilatata*, but occasionally restricted to the proximal part of the uterus. The size of spines is rather variable, 2-18 × 1-5 μm (length × width). Their number also differs significantly, from only a few scattered spines throughout the uterus to being numerous and almost completely filling the lumen. Depending on the peculiar means of attachment to the uterine wall, Kruger also distinguished three basic type of spines: smooth-based spines, stellate-based spines and spines attached in a uterine network;

iii) crystalline structures.

Several functions, summarised by Kruger (see also Grimaldi de Zio *et al.*, 1979), have been postulated for the uterine inclusions, namely slowing the passage of eggs through the uterus during shell formation, prevention of sperm reflux from the spermatheca during egg movement, and the secretion of certain substances from the globular structures for egg shell formation. Bleve-Zacheo *et al.* (1984, 1985) studied the ultrastructure of Z-differentiation in *Xiphinema* species and provided many details of its cell composition and arrangement, as well as its functional role.

Uterine eggs are frequently observed within the uterus. Their shape is ovoid, invariably longer than wide (Fig. 7.3G), sometimes elongate-ovoid (Fig. 7.3E) or rounded-ovoid (Fig. 7.3F). Egg size is very variable ranging from 40 to 300 long × 25 to 60 μm wide, even within the same species. Their number is also variable, but only a few (very often one) are usually found in the genital tract of a gravid female at the same time. The egg shell is secreted by the egg cell *per se* and may consist of three layers; an additional outer (fourth) layer composed of proteins is secreted by the uterus (Jairajpuri & Ahmad, 1992). The egg surface is very often (LM observations) smooth and regular, but an irregular surface is characteristic of *Aporcelinus* species and occasionally occurs in other genera (for example, *Laevides*, as noted by Shafqat *et al.*, 1994), and the presence of granules or short spines on the egg surface (echinate eggs, *cf.* Jairajpuri & Ahmad, 1992) has been very exceptionally mentioned in dorylaims, for example, in *Metaporcelaimus ovogranulosus* (see Álvarez-Ortega *et al.*, 2013; Fig. 7.3H).

According to Jairajpuri & Ahmad (1992), the transit of the oocyte/egg from the ovary through the oviduct and uterus to the exterior environment is most probably effected by a kneading action of the somatic muscles. This could be the reason for the usual absence of a muscular sheath around the genital tract. The presence of sphincters, as well as the narrow oviduct, most probably serves to prevent premature progress of the egg, since the muscles of the body undergo continuous contractions during locomotion.

As revealed by Bleve-Zacheo *et al.* (1993) for *Xiphinema diversicaudatum*, egg shell ultrastructure consists of three main layers, all formed at the level of the oviduct, plus two layers added to the surface of the egg, the inner secreted by the *pars dilatata oviductus* cells, and the outer secreted by the uterine cells. These layers are as follows: *i*) an outer vitelline layer, derived from the vitelline membrane, has a single membrane-like structure but becomes thickened by adhering particulate material (secreted by the oviduct cells) to its surface; *ii*) a median chitinous layer, of constant thickness, is made of chitin microfibrils surrounded by a protein coat that probably contains other substances; its main role may be to provide structural strength and resistance to stress in all directions; *iii*) the inner lipid layer, consisting of a lipoprotein complex whose protein component is minor, has a reticulate structure of variable thickness, being thicker at both poles; its role may be in the regulation of permeability; *iv*) during its passage through the *pars dilatata oviductus* a new layer made of phospholipids is added to the outer surface of the egg; and *v*) the outer layer of the egg shell, made of protein and mucopolysaccharides, is secreted by the glandular part of the uterus, including the Z-differentiation.

Vagina

Both uteri of di-ovarian species merge and join in a common vagina whereas the uterus of mono-ovarian species is followed by the vagina. The latter is a short tube-like section of the genital system leading to the body surface (Figs 7.5, 7.6). In general it extends inwards for one-third or one-half of the corresponding body diameter, but may occasionally be larger. The lumen is narrow, but its cross-shaped structure (Fig. 7.5.Aa, Da) enables great dilation.

De Ley *et al.* (1993) analysed and discussed in depth the morphology of the vagina[2], and proposed a new terminology that is very useful for

[2] The development of the vagina in *Caenorhabditis elegans* is regarded as a paradigm for organismal organogenesis, which has been a matter for detailed scrutiny (Sharma-Kishore *et al.*, 1999; Stemberg, 2005). Nevertheless, comparable studies are not available for any dorylaimid species.

its characterisation (Fig. 7.5). These authors distinguished three vaginal sections (Fig. 7.5B, C, E-G): proximal (*pars proximalis vaginae*), refringent (*pars refringens vaginae*) and distal (*pars distalis vaginae*). The *pars proximalis* (previously reported as the *vagina uterina*) is distinguished by its hyaline wall surrounded by a more or less conspicuous circular musculature (sphincter). It is described in terms of its length (along the lumen, from *pars refringens* to the proximal end of the hyaline wall), width (at level of the surrounding sphincter, but excluding the sphincter itself), and the shape of the outer margins of the wall (sigmoid, distally convergent, *etc.*). When present (it is lacking in many species), the *pars refringens* is a ring-like structure that, in lateral median view, appears as two sclerotisations (= sclerotised pieces). The *pars refringens* is the most variable part of the vagina and certainly the least prone to deformation. Its characterisation involves four aspects: *i*) absence/presence; *ii*) the shape of the sclerotisations (trapezoidal, drop-shaped, triangular, bar-shaped and semicircular); *iii*) size (length × width); and *iv*) the appearance of their proximal and distal surfaces (smooth, jagged, *etc.*) (Fig. 7.5H-N). (The reader is referred to De Ley *et al.* (1993) for an exhaustively rich explanation and characterisation of the *pars refringens*, including many actual examples.) Two comparatively large cells, one anterior and the other posterior, are present in a ventral position at the junction between the *pars proximalis* and the *pars refringens* in representatives of several dorylaimid genera such as *Crassolabium* (Fig. 7.6F), *Enchodelus* and *Nevadanema* (Fig. 7.3A, F). Finally, the *pars distalis* (previously reported as the *vagina vera*) is the region whose walls are formed by a continuation of the body cuticle. Depending on the nature (shape) of its walls in lateral view, the *pars distalis* can be straight, tapering, rounded, angular, *etc.* (Fig. 7.5O).

The oviposition of uterine eggs through the vagina and the vulva is facilitated by a complex system of muscles. Seinhorst & Kozlowska (1975) studied in great detail this system in *Longidorus* species, distinguished six groups of muscles associated with the vagina and the ovejector (Fig. 7.7), and proposed a precise terminology for their identification as follows: *i*) sphincter muscles around the ventral part of the vagina; *ii*) *dilatores vaginae*, attached to the lateral edges of the dorsal plate of the vagina and the body wall between the lateral chords and the dorsal body musculature; *iii*) muscles between the anterior and posterior edges of the dorsal plate and the body wall between lateral chords and dorsal body musculature; *iv*) *directores vaginae* (two pairs),

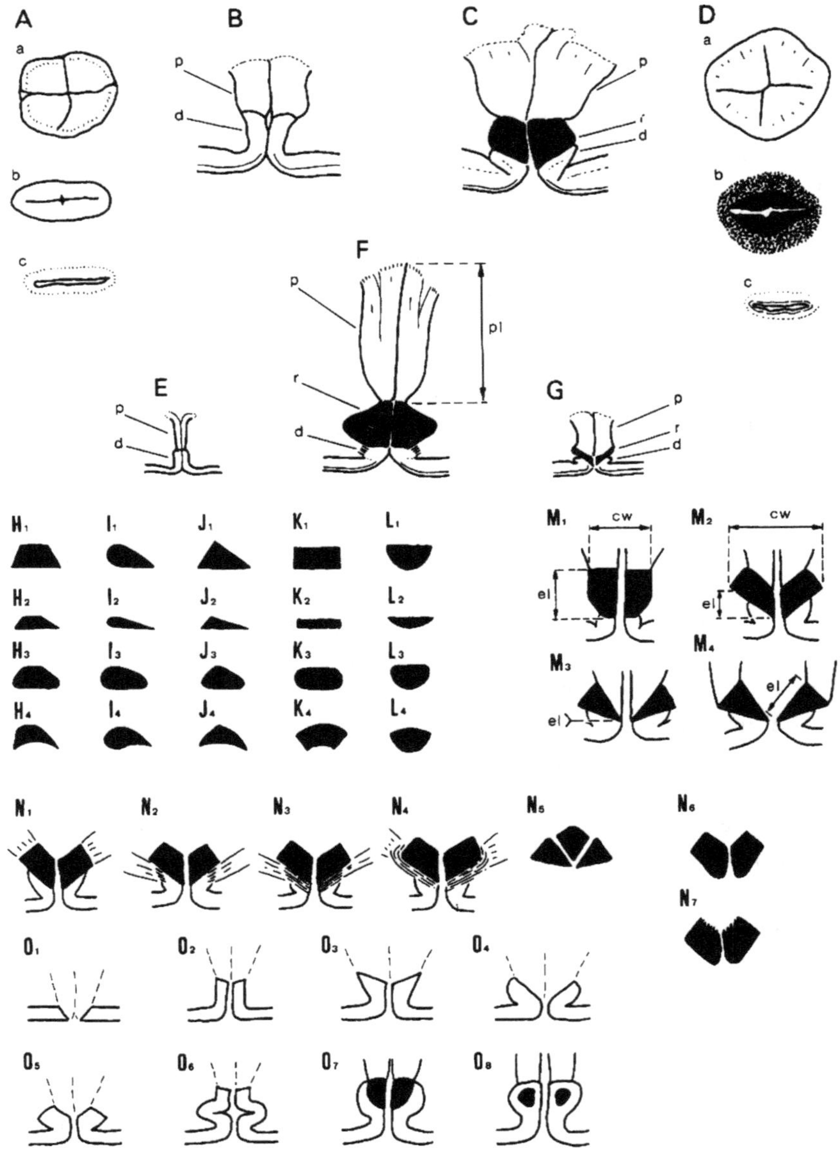

with both ends attached to the body wall between the lateral chords and the dorsal body musculature; *v*) dorsal *compressors* of the ovejector; and *vi*) muscles in the wall of the ovejector and uterus.

Vulva

The opening of the genital system on the body ventral surface (Fig. 7.8) is the vulva. It is usually a transverse slit (Fig. 7.8A, C), often somewhat oval (Fig. 7.8B), although a longitudinal vulva (Fig. 7.8D, E) is characteristic of a few genera, for example *Funaria* and *Paravulvus*, and even a pore-like shape has been described in particular species of some genera such us *Oriverutus*. Sometimes the vulva is located within a depression of the body surface, thus opening at the bottom of a small cavity. SEM techniques have provided additional data of the variability in vulva morphology. Two lips (anterior and posterior) are easily perceptible and well delimited in some species (Fig. 7.8A, B), and the existence of vulval flaps partially closing the opening is very distinct in other species (Fig. 7.8E). Vulva position is almost always more or less equatorial (at mid-body), but pre-equatorial (for example, *Tylencholaimellus* species) or post-equatorial (some members of *Tylencholaimus*) locations are also frequent. The *dilatatores vulvae* are specialised muscles extending from the lateral body wall to the internal side of the vulva. Cross-sections show these muscles to be close to the lateroventral somatic muscle fields, being attached between these muscles and the lateral chords (Coomans, 1965).

Fig. 7.5. *Detailed morphology of the vagina and its variability. A-G: Bi- or tripartite vagina. A: Optical sections in ventral view of vagina of* Carcharolaimus ramirezi *through* pars proximalis vaginae *(a),* pars distalis vaginae *(b) and* pars distalis vaginae*; B: Same in lateral view; C: Lateral view of vagina of* Aporcelaimellus *sp.; D: Optical sections in ventral view of vagina of* Aporcelaimellus *sp. through* pars proximalis vaginae *(a),* pars refringens vaginae *(b) and* pars distalis vaginae *(c); E: Vagina of* Aquatides *sp., lateral view; F: Vagina of* Labronema *cf.* mauritiense, *lateral view; G: Vagina of* Microdorylaimus diminutivus, *lateral view; H-N: Shapes and measurements of the* pars refringens vaginae*; O: Various states of the* pars distalis vaginae. *(Abbreviations: cw = combined width, d =* pars distalis vaginae, *el = extent along lumen, p =* pars proximalis vaginae, *pl = length of* pars proximalis vaginae, *r =* pars refringens vaginae.*) (After De Ley* et al. *(1993).* Bulletin de l'Institut Royal des Sciences Naturelles de Belgique *63, p. 16, Fig. 1; see original text for additional explanation.)*

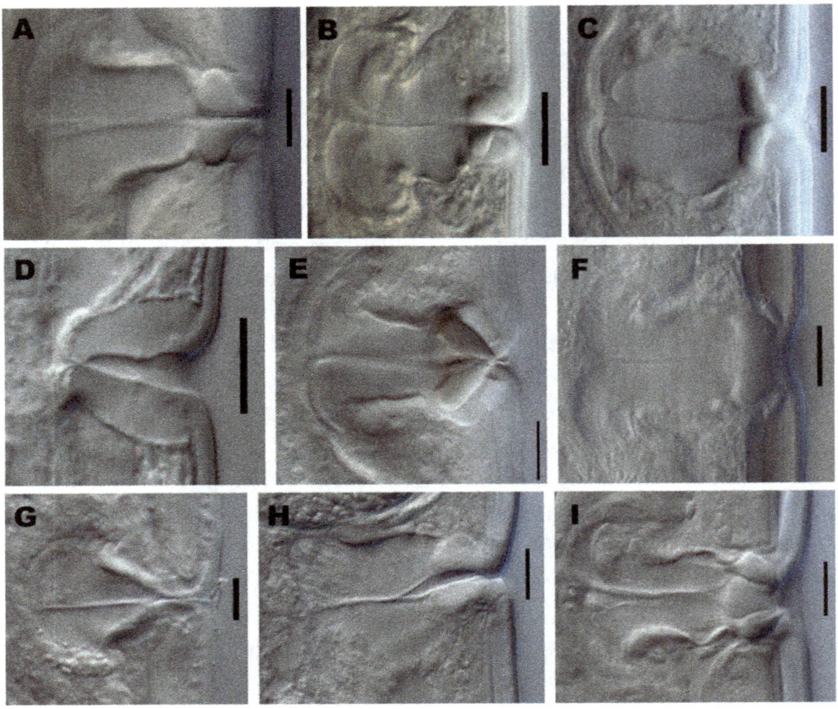

Fig. 7.6. *Vagina of selected dorylaimid species. A:* Aporcelaimellus obtusicaudatus *from Spain; B:* Aporcelaimellus salicinus *from Spain; C:* Aporcelaimellus rotundus *from Spain; D:* Adenolaimus orthus *from Iran; E:* Heterodorus veletensis *from Spain; F:* Crassolabium persicum *from Iran; G:* Proleptonchus *sp. from Spain; H:* Sectonema *sp. from Vietnam; I:* Labronema *sp. from Vietnam. (Scale bars: A-C, E, F, H, I = 10 μm; D, G = 5 μm.)*

Malformations or abnormalities

The appearance of morphological abnormalities affecting the female genital system of dorylaimid species has been repeatedly reported in the specialised literature (summarised in Table 7.1). These anomalies are probably caused by mutations or changes during the postnatal development. One kind of malformation is duplication, either total or partial. Thus, the duplication of the whole system, *i.e.*, specimens that bear four genital branches plus two vaginas and vulvae, has been noted in both free-living and plant-parasitic forms. The presence of three ovaries

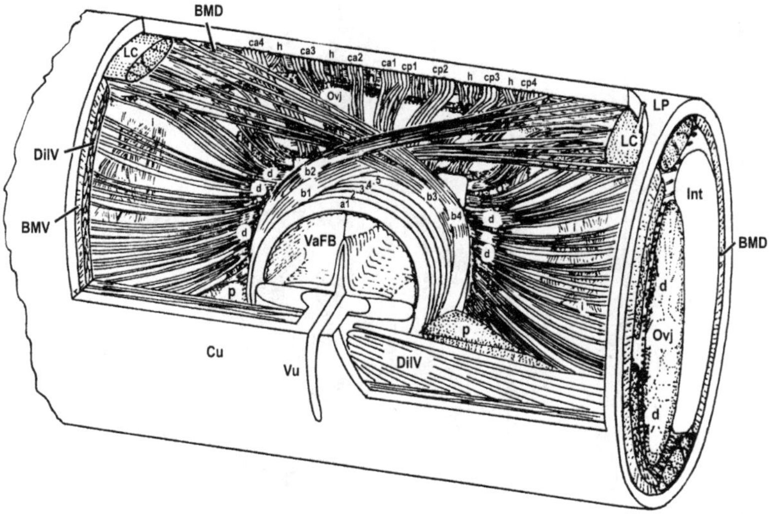

Fig. 7.7. *Musculature associated with the vagina and ovejector in* Longidorus caespiticola. *(Abbreviations:* a_1*-*a_5 *= sphincter muscles of vagina, BMD = dorsal body musculature, BMV = ventral body musculature,* b_1*-*b_4 = directores vaginae, *ca1-ca4 = anterior* dilatores vaginae, *cp1-cp4 = posterior* dilatores vaginae, *Cu = cuticle, d = ventral muscles between anterior or posterior edges of vaginal plate and body wall, DilV* = dilatores vulvae, *f = muscles in ovejector wall near anterior and posterior edges of vaginal plate, h =* dorsal compresores *of the ovejector, i = muscles of ovejector and uterus wall, Int = intestine, LC = lateral chord, LP = lateral pore, Ovj = ovejector, p = widened ventral chord, VaFB = vaginal 'filling bodies', Vu = vulva.) (After Seinhorst & Kozlowska (1975),* Nematologica *20(1974), p. 318, Fig. 9.)*

and oviducts – but only two uteri – due to a duplication of the anterior genital branch, was reported in *Xiphinema dentatum* by Radivojević (2005). The occurrence of specimens with two vulvae is not rare, but the presence of three vulvae has been reported only once, by Valocká & Sabová (1980).

The abnormal arrangement of the genital organs is another kind of anomaly. Thus, Brown & Coiro (1984) found a bivulval, didelphic-prodelphic, gravid female of *X. diversicaudatum*. The occurrence of didelphic, opisthodelphic females has been reported in two *Xiphinema* species (Arias, 1978; Cho & Robbins, 1990). The loss or reduction of some components of the system may also occur. Radivojević

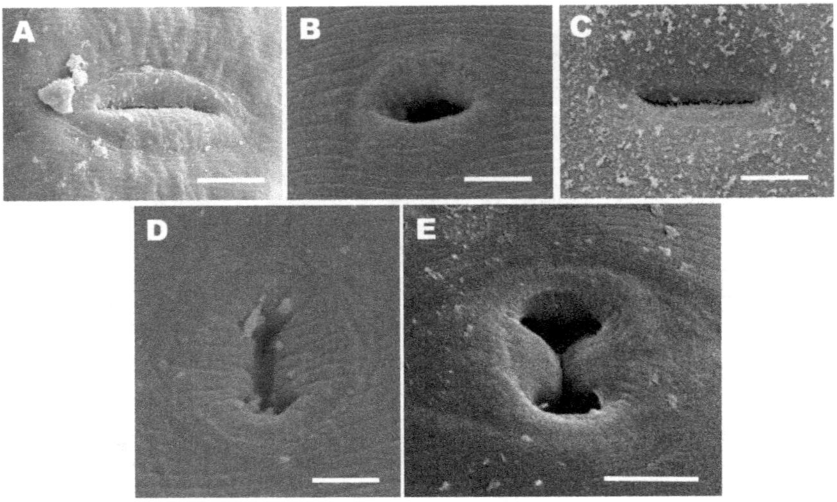

Fig. 7.8. *SEM pictures of vulva in selected dorylaimid species. A:* Aporcelaimellus obtusicaudatus *from Spain; B:* Oriverutus asaccatus *from Costa Rica; C:* Labronemella labiata *from Iran; D:* Labronema angeloi *from Spain; E:* Nevadanema nevadense *from Spain. (Scale bars: A, C, E = 5 μm; B, D = 2 μm.)*

(2005) found one female of *X. dentatum* that lacked both uteri and the vagina, its oviducts being connected by a sphincter-like structure. The reduction (shortening) and the complete loss of the anterior genital branch was also noted in two *Xiphinema* species (Arias, 1978; Cho & Robbins, 1990). A vulva-less specimen of *Labronema montanum*, which also lacked a vagina, was described by Peña-Santiago (2019).

The existence of hermaphrodite specimens is extremely infrequent and questionable. Luc (1961) described and illustrated a few females of *X. attorodorum* apparently bearing a testis instead of an ovary and interpreted it as a case of successive protandric hermaphroditism. Nevertheless, intersexuality, *i.e.*, the appearance of functional females possessing rudiments of male secondary sexual characters, but no trace of male gonads, has been reported by several authors (Table 7.1).

Table 7.1. *Malformations observed affecting the female genital system of dorylaims. (After Peña-Santiago, 2019; updated).*

Anomaly	Species	Habitat/Host plant	Country	Reference
Two duplicate systems	*Eudorylaimus* sp.		Spain	Peña-Santiago (1986)
	Aporcelaimellus obtusicaudatus			Geraert (1963)
	Xiphinema diversicaudatum	Mixed woodland	UK-Scotland	Brown & Coiro (1984)
Didelphic-opisthodelphic	*Xiphinema coxi coxi*	Alfalfa field	USA-Florida	Cho & Robbins (1990)
Loss of anterior genital branch	*Xiphinema coxi coxi*	Alfalfa field	USA-Florida	Cho & Robbins (1990)
Two vulvae	*Aporcelinus* sp.[1]	Rainforest	Costa Rica	Esquivel (2001)
	Californidorus cralleyi	?	USA	Robbins (1986)
	Longidorus artemisiae	*Artemisia* sp.	Russia	Rubtsova *et al.* (1999)
	Longidorus danuvii	*Salix alba*	Poland	Kornobis (2012)
	Longidorus elongatus		Russia	Robbins & Rubtsova (1996)
	Longidorus euonymus			Barsi (1994)
	Longidorus juvenilis	Vineyard	Slovenia	Sirca *et al.* (2007)
	Longidorus laevicapitatus	Coffee	São Tome	Lamberti *et al.* (1987)
	Longidorus sp.			Jairajpuri & Ahmad (1969)
	Mesodorylaimus bastiani			Loof (1969); Valocká & Sabová (1980)
	Nygolaimus sp.			Jairajpuri & Ahmad (1969)
	Xiphinema dentatum		Serbia	Radivojević (2005)
	Xiphinema diversicaudatum	Peach orchard	Czech Republic	Barsi (1994); Kumari & Decraemer (2009)

Table 7.1. *(Continued.)*

Anomaly	Species	Habitat/Host plant	Country	Reference
	Xiphinema index	Fig	Italy	Catalano (1991)
	Xiphinema turcicum	?	Former Yugoslavia	Radivojević (1991)
	Xiphinema vuittenezi	Vineyard	Italy	Coiro & Lamberti (1980); Barsi (1994)
		Apple orchard	Czech Republic	Kumari & Decraemer (2006, 2009)
Three vulvae	*Mesodorylaimus bastiani*			Valocká & Sabová (1980)
Vulva-less	*Labronema montanum*	Grassy soil	Spain	Peña-Santiago (2019)
Intersexual	*Aquatides thornei*	Freshwater-rice	India-Bihar	Jairajpuri *et al.* (1979)
	Enchodelus veletensis	Grass	Iran	Pedram *et al.* (2009)
	Funaria obtusa	?	Mexico	Goseco & Ferris (1973)
	Longidorus africanus	Avocado	Israel	Cohn & Mordechai (1968)
	Longidorus biformis	Elm, hackberry, maple	USA-Arkansas	Ye & Robbins (2004)
	Longidorus distinctus	Quince	Serbia	Krnjaic *et al.* (1999)
		Plum	Slovakia	Lisková (2007)
	Longidorus elongatus	?	UK-Scotland	Rasché & Boag (1981)
		?	USA	Robbins (1986)
	Longidorus helveticus	*Carpinus betulus*	Serbia	Barsi & De Luca (2005)
	Paralongidorus iranicus	Pine	Iran	Pedram *et al.* (2012)
	Tyleptus striatus	Sugarcane	India-Uttar Pradesh	Jairajpuri & Siddiqi (1964)
	Xiphinema ingens	Olive	Italy	Lamberti *et al.* (1975)
	Xiphinema insigne	Several	India-Uttar Pradesh	Bajaj & Jairajpuri (1977)

[1] Originally identified as *Eudorylaimus* sp., but the Figure 1 shows that the female belongs to a species of the genus *Aporcelinus*.

References

ABOUL-EID, H.Z. & COOMANS, A. (1966). Intersexuality in *Longidorus macrosoma*. *Nematologica* 12, 343-344.

ÁLVAREZ-ORTEGA, S., SUBBOTIN, S.A. & PEÑA-SANTIAGO, R. (2013). Morphological and molecular characterisation of Californian species of *Metaporcelaimus* Lordello, 1965 (Dorylaimida, Aporcelaimidae), with a new concept of the genus. *Nematology* 15, 251-278. DOI: 10.1163/15685411-00002674

ANDRÁSSY, I. (1991). The free-living nematode fauna of the Bátorliget Nature Reserve. In: Mahunka, S. (Ed.). *The Bátorliget Nature Reserve – after forty years*. Budapest, Hungary, Hungarian Natural History Museum, pp. 129-197.

ANDRÁSSY, I. (1996). Free-living nematodes in the Bükk mountains, Hungary. In: Mahunka, S. (Ed.). *The fauna of the Bükk National Park II*. Budapest, Hungary, Hungarian Natural History Museum, pp. 33-63.

ARIAS, M. (1978). Abnormal female gonad in *Xiphinema diversicaudatum* (Nematoda: Longidoridae). *Nematologia Mediterranea* 6, 231-233.

BAJAJ, H.K. & JAIRAJPURI, M.S. (1977). Variability within *Xiphinema insigne* populations from India. *Nematologica* 23, 33-46. DOI: 10.1163/187529277X00192

BARSI, L. (1994). Bivulval females of *Longidorus euonymus, Xiphinema diversicaudatum* and *X. vuittenezi* (Nematoda: Dorylaimida). *Nematologia Mediterranea* 17, 97-108.

BARSI, L. & DE LUCA, F. (2005). Morphological and molecular characterization of *Longidorus helveticus* (Nematoda: Dorylaimida) from Serbia. *Nematologia Mediterranea* 33, 41-49.

BLEVE-ZACHEO, M.T., CASTELLANO, M.A. & LAMBERTI, F. (1976). Preliminary studies on the ultrastructure of the female gonad of *Xiphinema index* and *X. mediterraneum* (Nematoda: Longidoridae). *Nematologia Mediterranea* 4, 41-55.

BLEVE-ZACHEO, M.T., ZACHEO, G., MELILLO, M.T. & LAMBERTI, F. (1984). On the structure of the Z pseudo-organ in *Xiphinema diversicaudatum*. *International Journal for Parasitology* 14, 405-413. DOI: 10.1016/0020-7519(84)90097-3

BLEVE-ZACHEO, M.T., ZACHEO, G. & LAMBERTI, F. (1985). Ultrastructure of the Z organ in *Xiphinema ifacolum*. *International Journal for Parasitology* 15, 453-461. DOI: 10.1016/0020-7519(85)90033-5

BLEVE-ZACHEO, M.T., MELILLO, M.T. & ZACHEO, G. (1993). Ultrastructural studies on the nematode *Xiphinema diversicaudatum*: egg shell formation. *Tissue and Cell* 25, 363-374. DOI: 10.1016/0040-8166(93)90078-Y

BROWN, D.J.F. & COIRO, M.I. (1984). A description of a bivulval, didelphic, prodelphic gravid *Xiphinema diversicaudatum* (Nematoda: Dorylaimida). *Nematologia Mediterranea* 12, 151-154.

CATALANO, L. (1991). A bivulval female of *Xiphinema index* (Nematoda: Dorylaimida). *Nematologia Mediterranea* 19, 223.

CHO, M.R. & ROBBINS, R.T. (1990). Observations of all postembryonic stages of *Xiphinema coxi coxi* (Nematoda: Longidoridae). *Journal of Nematology* 22, 69-78.

COHN, E. & MORDECHAI, M. (1968). A case of intersexuality and occurrence of males in *Longidorus africanus*. *Nematologica* 14, 591-592. DOI: 10.1163/187529268X00291

COIRO, M.I. & LAMBERTI, F. (1980). A female of *Xiphinema vuittenezi* with two vulvae. *Nematologia Mediterranea* 8, 87-89.

COOMANS, A. (1965). Structure of the female gonads in members of the Dorylaimina. *Nematologica* 10(1964), 601-622. DOI: 10.1163/187529264X00259

COOMANS, A. & CLAEYS, M. (1998). Structure of the female reproductive system of *Xiphinema americanum* (Nematoda: Longidoridae). *Fundamental and Applied Nematology* 21, 569-580.

COOMANS, A., CLAEYS, M. & HEYNS, J. (1992). Ultrastructure of the uterus of *Xiphinema pinoides* (Nematoda). *Fundamental and Applied Nematology* 15, 341-346.

COOMANS, A., HUYS, R., HEYNS, J. & LUC, M. (2001). Character analysis, phylogeny and biogeography of the genus *Xiphinema* Cobb, 1913 (Nematoda: Longidoridae). *Annales Sciences Zoologiques, Musée Royal de l'Afrique Centrale, Tervuren, Belgique* 287, 1-239.

DE LEY, P., LOOF, P.A.A. & COOMANS, A. (1993). Terrestrial nematodes from the Galápagos Archipelago II: Redescription of *Aporcelaimellus obtusicaudatus* (Bastian, 1865) Altherr, 1968, with review of similar species and a nomenclature for the vagina in Dorylaimida (Nematoda). *Bulletin de l'Institut Royal des Sciences Naturelles de Belgique* 63, 13-34.

ESQUIVEL, A. (2001). Una hembra de *Eudorylaimus* sp. con dos vulvas. *Nematropica* 31, 109-111.

GERAERT, E. (1963). *Aporcelaimus* female with two vulvae. *Nematologica* 9, 302-303. DOI: 10.1163/187529263X00520

GERAERT, E., GROOTAERT, P. & DECRAEMER, W. (1980). Structure of the female reproductive system in some Dorylaimida and Enoplida (Nematoda). *Nematologica* 26, 255-271. DOI: 10.1163/187529280X00143

GOSECO, C.G. & FERRIS, V.R. (1973). Intersexes of *Leptonchus obtusus* Thorne. *Journal of Nematology* 5, 226-228.

GRIMALDI DE ZIO, S., LAMBERTI, F. & MORONE DE LUCIA, M.R. (1975). The female gonad of *Longidorus africanus* Merny and the influence of the host on its development. *Nematologia Mediterranea* 3, 123-141.

GRIMALDI DE ZIO, S., D'ADDABBO-GALLO, M., LAMBERTI, F. & MORONE DE LUCIA, M.R. (1979). The "Z" differentiation in *Xiphinema*: a hypothesis of its function in relation to amphygony. *Nematologica* 25, 36-41. DOI: 10.1163/187529279X00352

JAIRAJPURI, M.S. & AHMAD, S. (1969). Record of bivulvar specimens of *Longidorus* sp. and *Nygolaimus* sp. *Science and Culture* 35, 492.

JAIRAJPURI, M. S. & AHMAD, W. (1992). *Dorylaimida. Free-living, predaceous and plant-parasitic nematodes*. Leiden, The Netherlands, Brill.

JAIRAJPURI, M.S. & SIDDIQI, A.H. (1964). Intersexuality in *Tyleptus striatus*. *Nematologica* 10, 182-183. DOI: 10.1163/187529264X00826

JAIRAJPURI, M.S., AHMAD, I. & AHMAD, M. (1979). Record of an intersex of *Aquatides thornei* with remarks on the phenomenon of intersexuality in nematodes. *Indian Journal of Nematology* 9, 177-181.

KORNOBIS, F.W. (2012). First record of *Longidorus danuvii* Barsi, Lamberti & De Luca, 2007 (Nematoda: Longidoridae) from Poland with description of pathologies of the lip region and reproductive system. *Helminthologia* 49, 104-107.

KRNJAIC, D., KRNJAIC, S., LAMBERTI, F. & AGOSTINELLI, A. (1999). New record of *Longidorus distinctus* Lamberti *et al.*, 1983 (Nematoda: Longidoridae) in Serbia. *Nematologia Mediterranea* 27, 57-61.

KRUGER, J.C. DE W. (1988). The uterine differentiation in *Xiphinema. Phytophylactica* 20, 233-251.

KRUGER, J.C. DE W. & HEYNS, J. (1988). Observations on the oviduct in *Xiphinema* (Nematoda: Dorylaimida). *Nematologica* 34, 493-498. DOI: 10.1163/002825988X00459

KRUGER, J.C. DE W. & HEYNS, J. (1990). Notes on the female reproductive system of *Xiphidorus* Monteiro, 1976 (Nematoda: Longidoridae). *Nematologia Mediterranea* 18, 47-51.

KUMARI, S. & DECRAEMER, W. (2006). A female of *Xiphinema vuittenezi* (Nematoda: Longidoridae) with two vulvae and abundant number of males in a soil sample. *Nematology* 8, 943-947. DOI: 10.1163/156854106779799196

KUMARI, S. & DECRAEMER, W. (2009). A single bivulval female of *Xiphinema diversicaudatum* and two bivulval females of *Xiphinema vuittenezi* (Nematoda: Longidoridae). *Helminthologia* 46, 250-252.

LAMBERTI, F., AGOSTINELLI, A. & NETO, E.S.S. (1987). A bivulval *Longidorus laevicapitatus* from Sao Tomé. *Nematologia Mediterranea* 15, 379-381.

LAMBERTI, F., BLEVE-ZACHEO, T. & MARTELLI, G.P. (1975). Un caso di intersesso in *Xiphinema ingens* Luc *et* Dalmasso (Nematoda, Longidoridae). *Nematologia Mediterranea* 3, 181-183.

LISKOVÁ, M. (2007). Morphometrics of females, juveniles and a hermaphrodite of *Longidorus distinctus* Lamberti *et al.*, 1983 (Nematoda: Longidoridae) from Slovakia. *Helminthologia* 4, 210-213.

LOOF, P.A.A. (1969). Taxonomy of some species of the genus *Mesodorylaimus* Andrássy, 1959. (Dorylaimoidea). *Nematologica* 15, 253-274. DOI: 10.1163/187529269X00687

LUC, M. (1961). Structure de la gonade femelle chez quelques espèces du genre *Xiphinema* Cobb, 1913 (Nematoda – Dorylaimoidea). *Nematologica* 6, 144-154. DOI: 10.1163/187529261X00405

LUC, M. (1973). Redescription de *Xiphinema hallei* Luc, 1958 et description de six nouvelles espèces de *Xiphinema* Cobb, 1893 (Nematoda: Dorylaimoidea). *Cahiers ORSTOM Série Biologie* 21, 45-65.

LUC, M. (1975). Taxonomy of *Xiphinema* Cobb, 1913. In: Lamberti, F. Taylor, C.E. & Seinhorst, J.W. (Eds). *Nematode vectors of plant viruses*. London, UK, Plenum Press, pp. 39-52.

MAGGENTI, A.R. (1981). *General nematology*. New York, NY, USA, Springer-Verlag New York, Inc.

PEDRAM, M., NIKNAM, G., GUERRERO, P., YE, W. & ROBBINS, R.T. (2009). Additional data on *Enchodelus veletensis* Guerrero, Liébanas & Peña-Santiago, 2007 (Dorylaimida: Nordiidae) from Iran, and phylogeny of the genus. *Nematology* 11, 217-229. DOI: 10.1163/156854109X429556

PEDRAM, M., POURJAM, E., NAMJOU, S., ATIGHI, M.R., CANTALAPIEDRA-NAVARRETE, C., LIÉBANAS, G., PALOMARES-RIUS, J.E. & CASTILLO, P. (2012). Molecular and morphological characterisation of *Paralongidorus iranicus* n. sp. and *P. bikanerensis* (Lal & Mathur, 1987) Siddiqi, Baujard & Mounport, 1993 (Nematoda: Longidoridae) from Iran. *Nematology* 14, 427-443. DOI: 10.1163/156854111X610272

PEÑA-SANTIAGO, R. (1986). *Eudorylaimus* sp. con sistema genital duplicado. *Nematologia Mediterranea* 14, 147-150.

PEÑA-SANTIAGO, R. (2019). A very rare malformation affecting the female genital system of one *Labronema* specimen (Dorylaimida, Dorylaimidae). *Journal of Nematology* 51, e2019-17. DOI: 10.21307/jofnem-2019-017

RADIVOJEVIĆ, M. (1991). *Xiphinema turcicum* Luc & Dalmasso, 1964 first stage juvenile with double spear, and female with two vulvae. *Revue de Nématologie* 14, 321-322.

RADIVOJEVIĆ, M. (2005). Genital organ anomalies in *Xiphinema dentatum* (Nematoda: Longidoridae). *Nematology* 7, 295-299. DOI: 10.1163/1568541054879601

RASCHKÉ, I.E. & BOAG, B. (1981). Morphological variation and distribution frequency of hermaphrodite *Longidorus elongatus*. *Revue de Nématologie* 4, 283-285.

ROBBINS, R.T. (1986). Observations of *Californidorus cralleyi* with two vaginas and an intersex of *Longidorus elongatus*. *Journal of Nematology* 18, 640.

ROBBINS, R.T. & RUBTSOVA, T.V. (1996). A specimen of *Longidorus elongatus* with two vulvas from Russia. *Nematropica* 26, 305.

RUBTSOVA, T.V., CHIZOV, V.N. & SUBBOTIN, S.A. (1999). *Longidorus artemisiae* sp. n. (Nematoda: Longidoridae) from roots of *Artemisia* sp., Rostov region. *Russian Journal of Nematology* 7, 33-38.

SEINHORST, J.W. & KOZLOWSKA, J. (1975). The muscles of the vagina and the ovejector in *Longidorus caespiticola*. *Nematologica* 20(1974), 308-322. DOI: 10.1163/187529274X00357

SHAFQAT, S., AHMAD, I. & AHMAD, W. (1994). SEM morphology and embryology of *Laevides symmetricus* (Nygolaimina: Nygolaimidae). *Nematologica* 40, 534-540. DOI: 10.1163/003525994X00409

SHARMA-KISHORE, R., WHITE, J.G., SOUTHGATE, E. & PODBILEWICZ, B. (1999). Formation of the vulva in *Caenorhabditis elegans*: a paradigm for organogenesis. *Development* 126, 691-699.

SIRCA, S., STARE, B.G., PLESKO, I.M., MARN, M.V. & UREK, G. (2007). First record of *Longidorus juvenilis* and *L. leptocephalus* (Nematoda: Dorylaimida) in Slovenia and their morphometrical and ribosomal DNA sequence analysis. *Russian Journal of Nematology* 15, 1-8.

STENBERG, P.W. (2005). Vulval development. In: The *C. elegans* Research Community (Ed.). *WormBook*. DOI: 10.1895/wormbook.1.6.1.

VALOCKÁ, B. & SABOVÁ, M. (1980). Morphological anomalies in the reproductive system of *Mesodorylaimus bastiani* (Bütschli, 1873) Andrássy, 1959. *Helminthologia* 17, 141-144.

VAN DE VELDE, M.C. & COOMANS, A. (1988). Electron microscopy of the germ cells and the ovarian wall in *Xiphinema* (Nematoda). *Tissue and Cell* 20, 881-890. DOI: 10.1016/0040-8166(88)90029-8

VAN DE VELDE, M.C., COOMANS, A., HEYNS, J. & CLAEYS, M. (1990a). Ultrastructure of the female reproductive system of *Xiphinema meridianum* (Nematoda). *Revue de Nématologie* 13, 211-223.

VAN DE VELDE, M.C., COOMANS, A., HEYNS, J., CLAEYS, M. & HUTSEBAUT, M. (1990b). Ultrastructure of the female gonoduct of *Xiphinema theresiae* (Nematoda). *Revue de Nématologie* 13, 449-461.

YE, W. & ROBBINS, R.T. (2004). *Longidorus biformis* n. sp. and *L. glycines* n. sp. (Nematoda: Longidoridae): Two amphimictic species from Arkansas. *Journal of Nematology* 36, 1-13.

Chapter 8

Male genital system

General concept

The male genital system of dorylaims invariably consists of two gonads (testes), a condition termed diorchic[1], which merge in a common duct that connects the rectum, forming together the cloaca, the latter leading to and opening on the body surface as the cloacal aperture. Other organs and structures serve as accessory elements: copulatory apparatus (spicules, lateral guiding pieces, gubernaculum, and associated musculature), and copulatory muscles, glands, and genital papillae (supplements). Aboul-Eid (1969) and Jairajpuri & Ahmad (1992) provided the most detailed descriptions of the system available so far, and their ideas are herein mainly followed and summarised. The study of the male genital system in dorylaims has not received much attention as it shows minimum variation in its basic morphology, a fact that contrasts significantly with those observed in the female system. The accessory elements of the system, however, display an extraordinary range of shapes, sizes and (if present) number, and deserve deeper analysis and discussion.

Testes

A pair of opposed testes are always observed in dorylaims. The distal end of the anterior testis is anteriorly directed (Fig. 8.1A, D), whereas the distal end of the posterior testis is posteriorly directed (Fig. 8.1B). Both testes are nearly identical in size. Under LM, each testis appears as an elongate (several times as long as wide) structure with a thin wall made of epithelial layer, which is more distinct at its proximal half and nearly indistinguishable at the distal half. The epithelial layer may be

[1] Totally exceptional, and certainly an abnormality, is the existence of one (monorchic condition) or three (triorchic condition) testes as reported in *Xiphinema* species by Bajaj & Jairajpuri (1978).

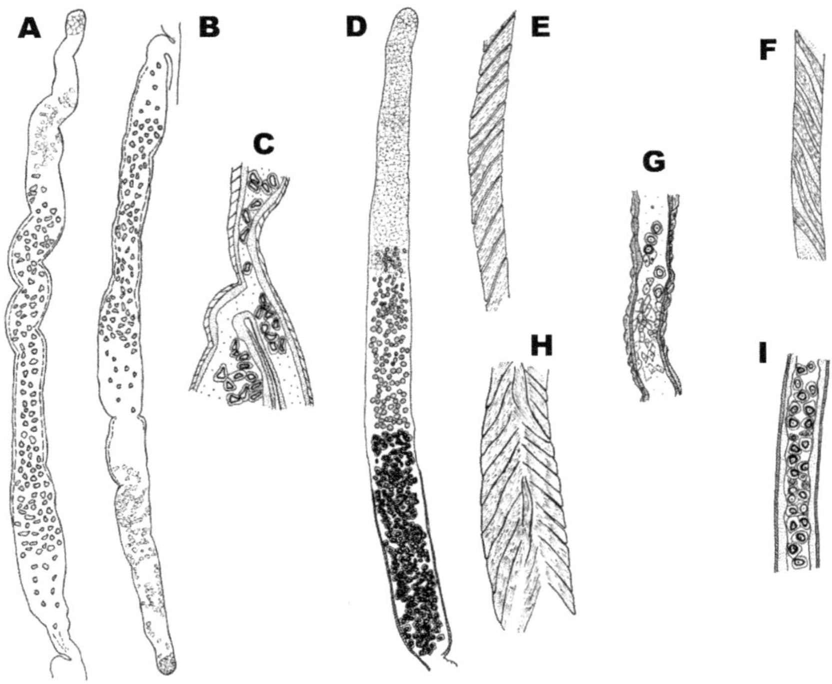

Fig. 8.1. *Testes and deferent canal. A, D: Anterior testis; B: Posterior testis; C: Anterior and posterior testes merging into deferent canal; E: Musculature of distal region in anterior testis; H: Same in proximal region; F: Musculature of deferent canal; G: Longitudinal section of testis showing epithelial layer and muscles; I: Sperm cells inside the deferent canal. A-C:* Longidorus macrosoma*; D-I:* Longidorus caespiticola*. (After Aboul-Eid (1969),* Nematologica *15, pp. 442, 443, Figs 3, 4.)*

covered by prominent oblique muscles (Fig. 8.1E, G, H). Two regions are invariably distinguishable in the testes, namely the germinative or multiplication zone and the maturation or differentiation zone.

In size and shape, sperm cells[2] are quite variable (Fig. 8.2), but very often they appear as fusiform (spindle-shaped) cells, less frequently more rounded or oval in shape. Their origin is the result of a complex process of differentiation and ripening that takes place in their passage through the testis. The distal part of germinative region contains

[2] Since the male gametes of dorylaims lack a tail, the term 'sperm cell' should be used instead of 'spermatozoa' (*cf.* Jairajpuri & Ahmad, 1992: p. 45).

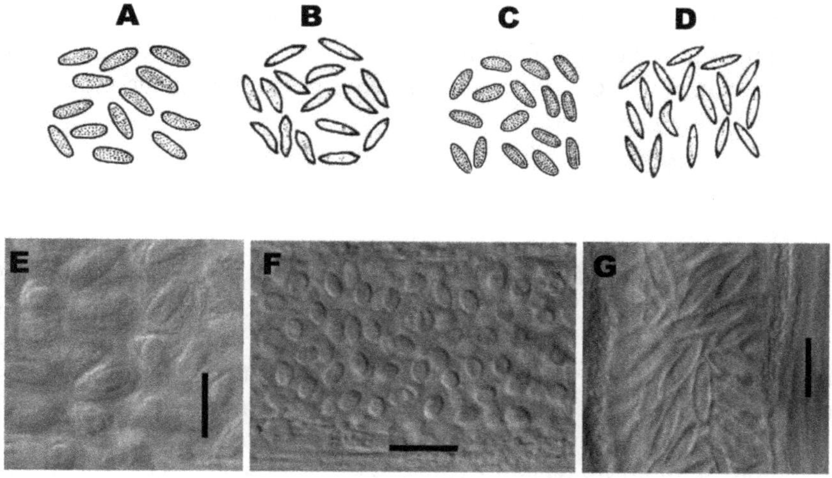

Fig. 8.2. *Sperm cells in selected dorylaims. A-D:* Axonchium *species. A:* A. elegans*; B:* A. sabulum*; C:* A. saccatum*; D:* A. transkeiense *(after Nair (1973),* Nematologica *18(1972). pp. 524, 534, 539 and 546, respectively); E:* Sectonema septentrionale *from Spain; F:* Aporcelinus itecrii *from Costa Rica; G:* Labronema montanum *from Spain. (Scale bars = 10 μm.)*

abundant germ cells (spermatogonia) with prominent nuclei, but lacking perceptible cell boundaries. At the proximal part of the germinative region the cells become larger and bear a well-defined wall. The distal part of the maturation region includes a large number of squarish or hexagonal cells (spermatocytes) with prominent nuclei, each of them resulting in four spermatids. The latter are small rounded cells with smaller nuclei than the spermatocytes and, after maturation, culminate in the formation of sperm cells.

Bacceti *et al.* (1983) stated that nematode sperm cells are characterised by the absence of both a flagellum and an acrosome, and that those of Dorylaimida (represented by the longidorid species *X. diversicaudatum* in their study) lack spheroidal membranous vesicles in the cell periphery, which are present in many other nematode taxa. More recently, Yushin (2010) described the ultrastructure of sperm cells of *Paractinolaimus microdentatus*, a member of the family Actinolaimidae, and concluded that these sperm cells show a very special internal structure that distinguishes them from those found in other nematode taxa and which may be regarded as a distinctive character of the order Dorylaimida. Each sperm cell contains a central complex of a nucleus and mitochondria, which is surrounded by electron-light cytoplasm containing numerous bundles of filaments and microtubule-like fibres 16-18 nm in diameter, with no inclusion resembling the membranous organelles found in other nematode taxa.

Genital tract

Both testes merge at their proximal ends and lead to the genital tract (Fig. 8.1C). This is a long tubular structure that lies parallel and ventral to the intestine plus prerectum and connects to the rectum. It is especially distinct in well-preserved specimens, but its detailed morphology is difficult to observe. It consists of two sections, namely a deferent canal (*vas deferens*) and the ejaculatory duct. The junction between both testes is simple, with no sphincter or valve-like structure, but the muscles that cover the two testes become more strongly developed in that region (compare Figs 8.1E and 8.1H).

The deferent canal is lined with epithelial cells and its walls may be highly granular (Fig. 8.1I). Oblique muscles, similar to those found over the testes, are also present around the deferent canal and may serve to push the sperm cells through its lumen (Fig. 8.1F). The transition from deferent canal to ejaculatory duct is hardly perceptible, but the former usually narrows before leading into the latter.

The walls of the ejaculatory duct are thick, made of epithelial cells and packed with granules, and the lumen is narrow but filled with sperm cells some time before copulation. The ejaculatory duct narrows gradually as it proceeds to join the rectum to form the cloaca a short distance posterior to the prerectum-rectum junction.

The cloaca is a short and narrow tube-like structure, serving as a common evacuation duct for both the digestive and the male reproductive systems. Its dorsal wall bulges to form sacs or pouches that accommodate the copulatory apparatus. The cloaca wall is lined with cuticle. It terminates on the body surface at the cloacal aperture, whose morphology is quite similar to the anus of the female.

Spicules

Always paired, the spicules are the major components of the copulatory apparatus. Their nature is similar to that of the cuticle, but they become heavily sclerotised and refringent structures. Embryologically, they derive from a special group of cells, termed *spicula primordia*, located in the wall of the cloaca, which form a dorsal sac or pouch (see above) along with the spicules. The sac is lined with cuticle that is continuous with the cuticle of the cloaca (Jairajpuri & Ahmad, 1992).

For practical (descriptive and comparative) purposes, and based on Andrássy's (1997) scheme, a definition of the constitutive parts or elements of the spicule was provided by Peña-Santiago *et al.* (2014) as follows. As observed in lateral view, these components may be usually distinguishable (Fig. 8.3A):

i) Dorsal contour, often regularly convex, with no special differentiation.
ii) Ventral contour, always (more or less, often distinctly) concave, usually bearing a variably prominent hump, located at *ca* the middle of the spicule total length and followed by an also variably marked, adjacent hollow.
iii) Median piece, more or less developed (only exceptionally absent), often occupying most of the inner core, occasionally reaching the spicule anterior end and in general ending at its posterior tip.
iv) Head (*capitulum*), the spicule anterior part, which, for practical reasons, is considered to run from the spicule anterior end to the starting point of the median piece.
v) Blade (*lamina*), the remaining part of the spicule, excluding the head.

More special elements or differentiations are known to occur but they have been recorded in very particular taxa. For example, Andrássy (1986) characterised the genus *Calcaridorylaimus* by having a distinct dorsal spur at the distal end of each spicule.

Unfortunately, the spicule description is often not very detailed, even when new taxa are proposed; nevertheless, it is almost always acceptably drawn or illustrated and useful for comparative purposes. The shape and the size of spicules are very constant within the same species and, for bisexual species, provide good diagnostic characters. Conversely, the differences between species are often very significant and of high value for their separation (Fig. 8.3B), even if the latter belong to the same genus (Fig. 8.4). Some spicule features to take into consideration are listed in the following. The presence/absence of a distinct head as well as its morphology (small or long compared to the blade, straight

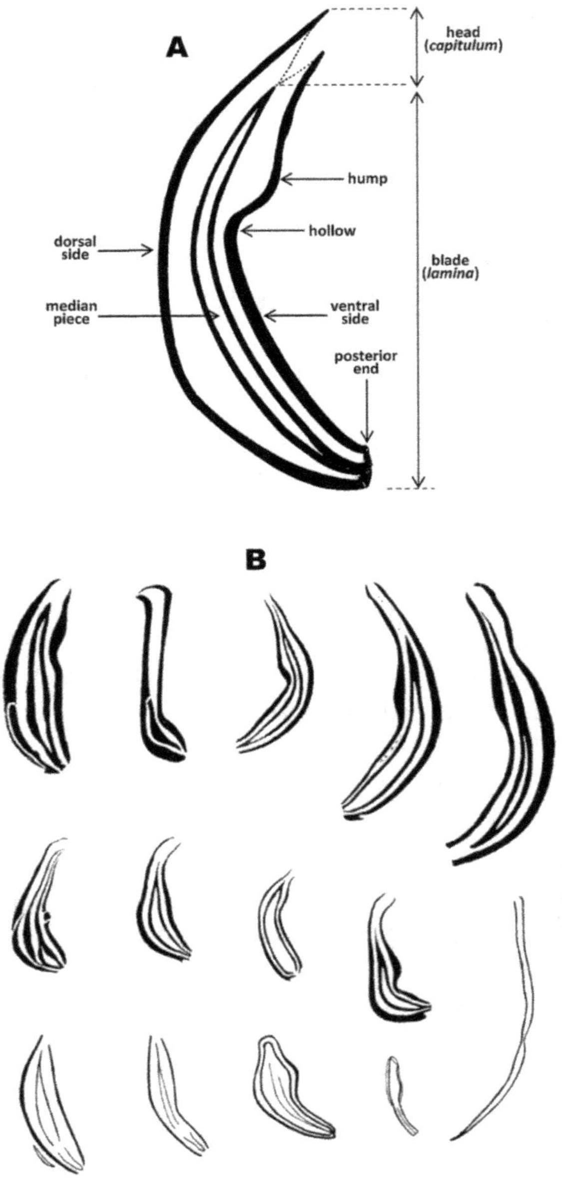

Fig. 8.3. *Spicule form in dorylaims. A: General morphology, showing most relevant parts and regions (after Peña-Santiago* et al. *(2014),* Nematology *16, p. 1093, Fig. 2); B: Variation in shape and size (after Jairajpuri & Ahmad (1992),* Dorylaimida, *p. 48, Fig. 22 in part).*

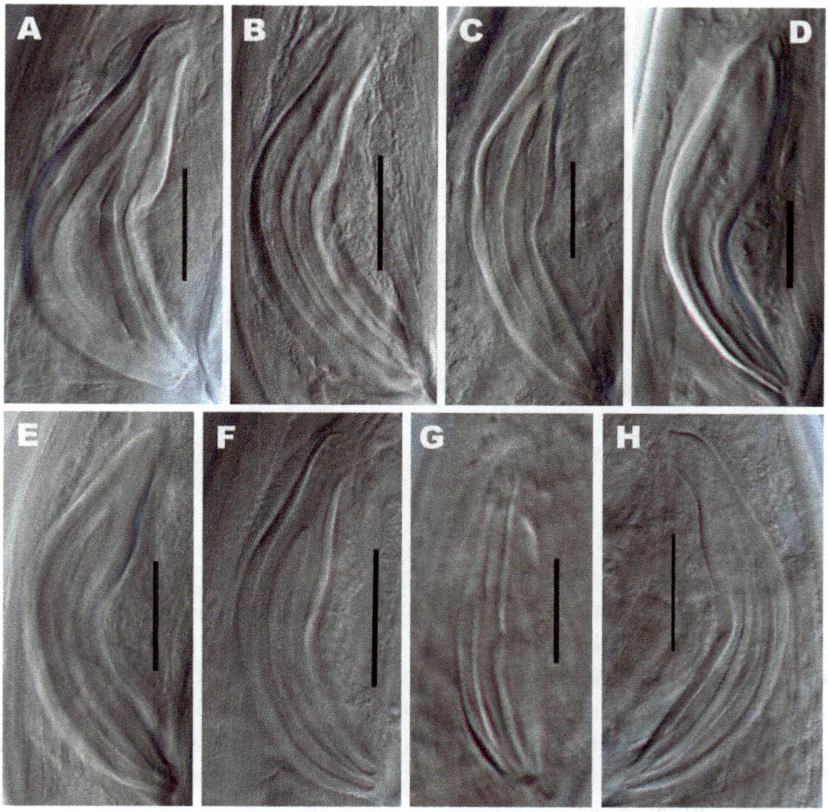

Fig. 8.4. *Spicules of selected dorylaimid species. A:* Sectonema *sp. from Vietnam; B:* Labronemella labiata *from Iran; C:* Aporcelaimellus obtusicaudatus *from Spain; D:* Belondira *sp. from Vietnam; E:* Crassolabium persicum *from Iran; F:* Enchodelus longispiculus *from Iran; G:* Morasia *sp. from Vietnam; H:* Stenodorylaimus textilis *from the USA. (Scale bars: A-C, E, F, H = 20 μm; D, G = 10 μm.)*

or curved, open or closed) have been often omitted or overlooked in descriptions, but illustrations may provide these details. The curvature of spicules is highly variable, ranging from nearly straight forms to others displaying strong (right) angles. The presence/absence of ventral hump, its development and position are also highly variable. The shape of the internal median piece and its extension, for example, if it reaches the anterior (distal) and/or the posterior (proximal) ends, are also important.

The morphology of the posterior end of blade (rounded or acute, open or closed, *etc.*) provides useful information.

Finally, morphometrics are very important and fundamental for species characterisation. Peña-Santiago *et al.* (2014) proposed a series of morphometrics for this purpose, either measurements (spicule total length at both the arc and the chord, head length and width, spicule maximum width, position of the ventral hump, spicule width at its posterior tip, length and width of the median piece, and spicule curvature) or ratios and percentages (spicule total length/spicule maximum width, spicule total length/body diameter at cloacal aperture level, median piece length/median piece maximum width, head length as percentage of spicule total length, position of ventral hump as percentage of spicule total length, and median piece width as percentage of spicule maximum width).

Lateral guiding pieces

Two small structures, each located on either side of the spicule, probably assist in the protrusion of the spicules and are hence termed the lateral guiding pieces. Only exceptionally absent, the lateral guiding pieces are much smaller than the spicules and are located near their posterior end. In spite of their small size, the lateral guiding pieces are highly variable in morphology. They are often more or less cylindrical, straight and moderately long (3-6 times as long as wide), but less frequently curved ventrad with an acute or bifurcate (bifid) tip, *etc.* (Fig. 8.5). There are separate sets of retractor muscles, which serve to retract them into the body (Jairajpuri & Ahmad, 1992). The shape and size of the lateral guiding pieces show little intraspecific variability and may be useful for species characterisation but unfortunately they are not always described with accuracy.

Gubernaculum

This element of the copulatory apparatus, so frequent in many nematode taxa, is almost entirely absent in dorylaims. Only a few genera, most of them belonging to the suborder Nygolaimina, for example, *Aquatides* and *Laevides*, bear such a structure, but it is never very well developed and takes the form of a simple, small, sometimes slightly

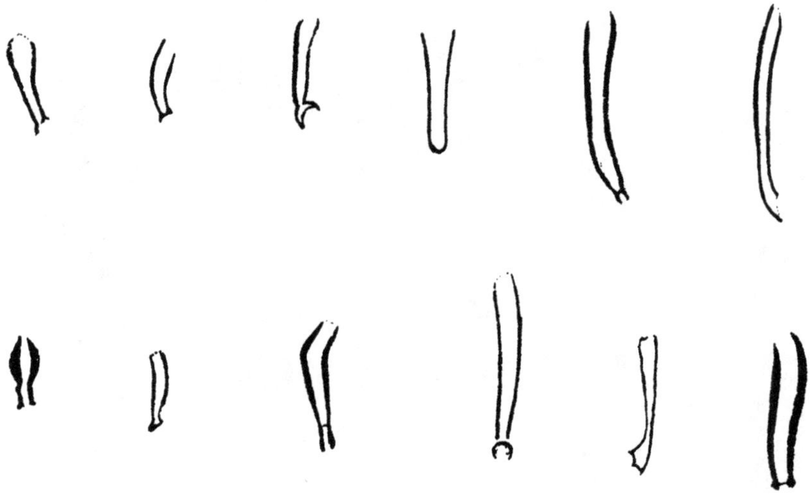

Fig. 8.5. *Variation in shape and size of the lateral guiding pieces in dorylaims (after Jairajpuri & Ahmad (1992), Dorylaimida, p. 48, Fig. 22 in part).*

arcuate, plate-like structure. It is a thickening of the dorsal wall of the cloacal pouch and also serves to guide the movement of spicules during copulation (Jairajpuri & Ahmad, 1992). Because of its rare occurrence, the gubernaculum is useful for the identification of the dorylaims when present.

Genital papillae (supplements) and other papillae

Dorylaimid males are characterised by having a variable number of copulatory or genital papillae, often also termed supplements. These papillae have a papilliform or mammiform aspect, usually appearing as small button-like structures, more or less protruding from the body surface. There is one pair of ad- or precloacal supplements and a series of individual (very exceptionally paired) supplements (Fig. 8.6A). The ad- or precloacal pair is situated close together, in a ventral position and at a short distance anterior to the cloacal aperture (Fig. 8.6A-C). The presence of this pair of supplements is an exclusive feature (an autapomorphic condition) of dorylaims. The series of individual supplements is located along the mid-ventral line of the posterior body region, and hence they are referred to as the ventromedian supplements.

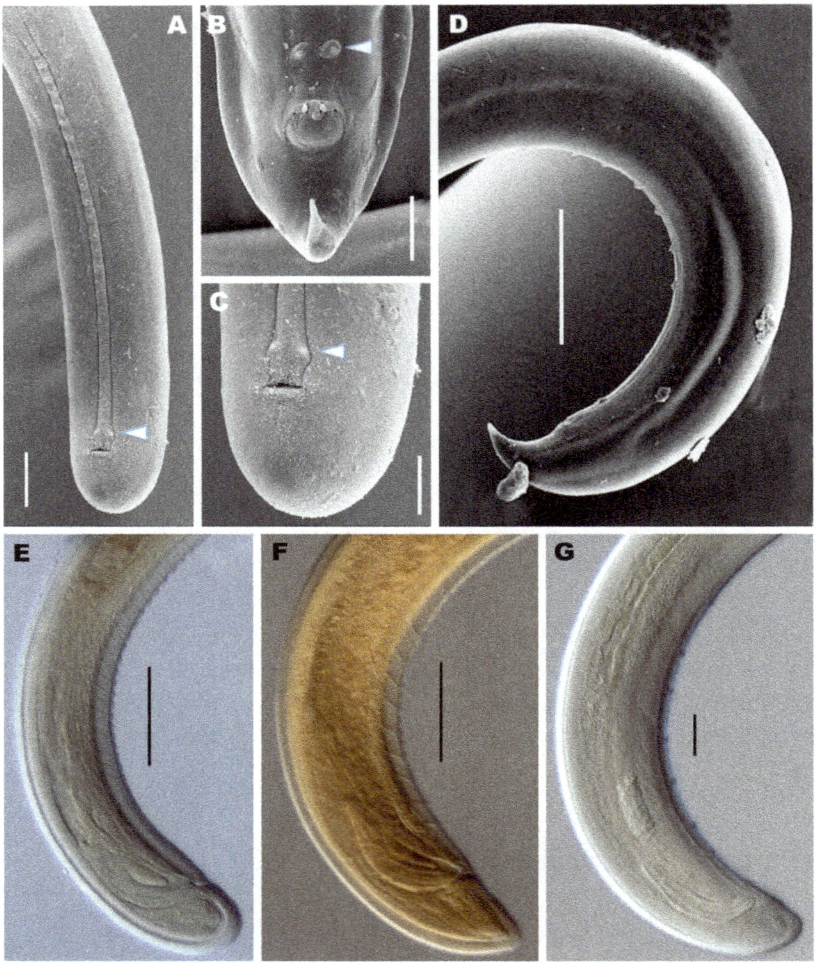

Fig. 8.6. *Male genital papillae in dorylaims. A: Posterior body region of* Labronemella labiata *from Iran, showing the genital papillae in ventral view; B, C: Precloacal pair (arrow head) of genital papillae in* Rhyssocolpus iuventutis *from Spain (B) and* Labronemella labiata *from Iran (C), ventral view; D: Posterior body region (lateral view) of* Rhyssocolpus iuventutis *from Spain, showing location of male genital papillae; E-G: Male posterior body region of selected species, showing contiguous ventromedian supplements in* Labronema neopacificum *from Vietnam (E), closely spaced ventromedian supplements in* Stenodorylaimus textilis *from the USA (F), and widely spaced ventromedian supplements in* Enchodelus longispiculus *from Romania. (Scale bars: A = 20 μm; B, C = 10 μm; D-F = 50 μm.)*

Ventromedian supplements are extremely variable among dorylaims and their number and arrangement are very useful taxonomical features for the identification of bisexual species. The number ranges from zero to more than 100, but seldom fewer than three or more than 30. Their arrangement displays an interesting variation among the taxa. On the one hand, the supplements may be contiguous or spaced. In the first case, the genital papillae are arranged close together with no space in between (Fig. 8.6E) whereas, in the second case, each papilla is separated from its adjacent one(s) by a perceptible and measurable, but highly variable, distance (Fig. 8.6D, F, G). A very special case is represented by a few genera in which the supplements are arranged in few, very compact and separated groups termed fascicles (Fig. 8.7C). On the other hand, the position of the ventromedian series with respect to the pre-cloacal pair may present two basic patterns. The ventromedian series is very often widely separated from the adcloacal pair, the most posterior supplement situated outside the range of the retracted spicules, forming a so-called precloacal space or hiatus in the sequence (Fig. 8.6A, D, E). In some groups the ventromedian series is not so distant from the adcloacal pair, with the most(s) posterior supplement(s) lying within the range of spicules. In this type the hiatus is lacking (Fig. 8.6F, G).

The genital papillae appear to be sensory as well as glandular in nature. A nerve ending can be seen protruding slightly from the centre of each papilla. The frequent presence of an exudation of some kind around the supplements indicates that they function as organs of secretion during copulation (Jairajpuri & Ahmad, 1992).

Dorylaimid males very often bear other papillae or pores, located subventrally, in the region of the supplements. Their number is highly variable and they may even be absent. Their precise function is not known.

Specialised musculature

The posterior body region of males bears a complex musculature (Fig. 8.7) that is mainly related to copulation. Das (1963) and Aboul-Eid (1969) studied this musculature in *Aquatides aquaticus* and *Longidorus macrosoma*, respectively, and their main findings are compiled and summarised below.

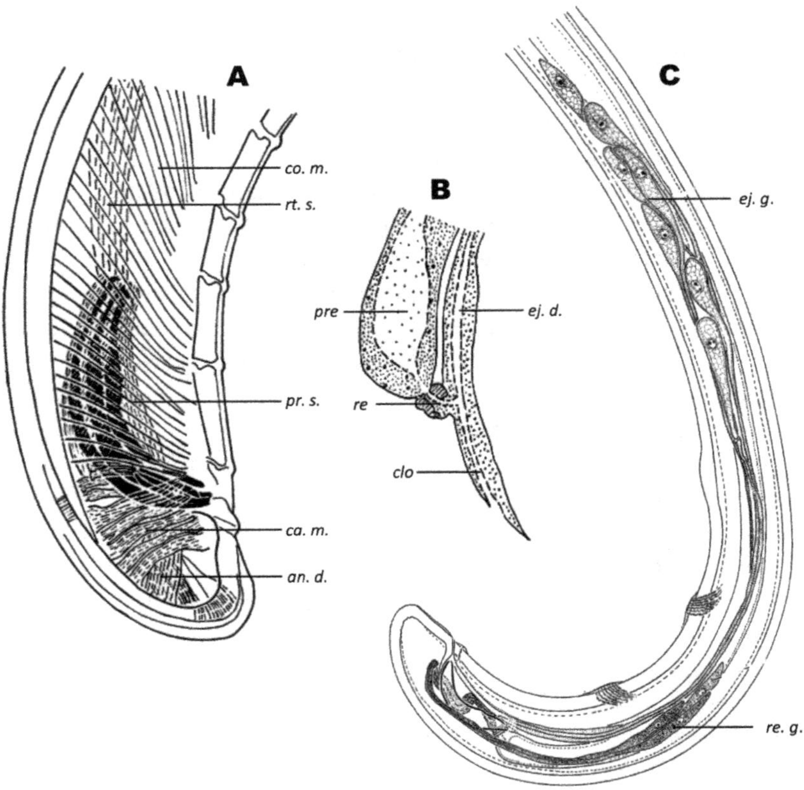

Fig. 8.7. *Posterior body region of dorylaimid males showing musculature (A) and glands (C) associated with male genital system, and junction of digestive and genital system (B). (Abbreviations:* an. d. = *anal dilator,* ca. m. = *caudal copulatory muscles,* clo = *cloaca,* co. m. = *copulatory muscles* ej. d. = *ejaculatory duct,* ej. g. = *ejaculatory glands,* pre = *prerectum,* pr. s. = protractores spiculae, re = *rectum,* re. g. = *rectal glands,* rt. s. = retractores spiculae.*) (A, B: After Aboul-Eid (1969),* Nematologica *15, pp. 445, 447, Figs 5 in part; 6C; C: After Coomans & Loof (1986),* Revue de Nématologie *9, p. 262, Fig. 1C, modified.)*

A series of muscle bands, termed the copulatory muscles, are present in the pre-cloacal region and extend over a wide area, for example, in *Aquatides aquaticus* they occur over about one-fifth of the total body length, from the position of the adcloacal pair of supplements to a short distance anterior to the most anterior ventromedian supplements. The number of muscle bands varies within and between species. The bands run obliquely from a dorsolateral position to a subventral position on both sides of the body and they take part in, or facilitate, copulation. Their probable function is to cause the curvature of the male posterior region in order to embrace the female body firmly during copulation.

Two groups of muscles, termed caudal copulatory muscles, extend more or less transversely from each laterodorsal body wall, the most anteriormost muscles attaching to the posterior lip of the cloacal opening. Together with the copulatory muscles, they ensure a greater curvature of the posterior region of the male during copulation.

The anal dilator muscle extends posteriorly from the distal cloacal wall to the laterodorsal wall of the tail.

From a resting position, the posterior ends of the spicules can be protruded from their pouches through the cloacal-spicular orifice, which lies very close to the cloacal opening. Two sets of muscles facilitate the movement of the spicules during copulation, of special importance being the protractor and the retractor muscles, both sets being attached to the distal (anteriormost) end of the spicules. The protractor muscles are connected to the anteriormost end of the spicules and run to the dorsal wall at the tip of the tail. The retractor muscles are located anterior to the spicules, and run from their anteriormost end to the dorsal side of the body wall. The protractor muscles extend posteriad and their contraction results in the protrusion of spicules, whereas the retractor ones extend anteriad and facilitate retraction.

The muscles associated with the guiding pieces are retractor muscles that extend from the laterodorsal body wall to the sheath of each guiding piece. Contraction of these muscles serves to retract the protruding guiding pieces to the interior of the body.

Glands associated with male genital system

Coomans & Loof (1986), in a detailed study, described two sets of gland cells associated with the male genital system in *Parastomachoglossa costata*, namely the ejaculatory glands and the rectal glands. In addition, the authors discussed the phylogenetic importance of these gland cells. The following text is based on their paper as no relevant later contribution about these structures is available.

The ejaculatory glands usually consist of eight (sometimes fewer) cells, the bodies of which lie either on both sides of the intestine or partly to one lateral side of the intestine and ventrally from it, so that all of them can be observed from one side (Fig. 8.7C). Depending on the length of the prerectum the cells may lie partly anterior to the intestine-prerectum junction. The cell bodies lead to long ducts running on both sides of the prerectum and/or *vas deferens*. The ducts merge with the ventral wall of the *vas deferens* and then open into its lumen.

The rectal glands consist of three pairs of cells located dorsolaterally or laterally from the prerectum. Their ducts first run parallel to the prerectum, then run dorsolaterally above and along the spicules, curve inwards opposite or just posterior to the cloacal opening, run anteriorly between the spicules into the dorsal wall of the cloaca, and then open into the cloaca just posterior to the junction between the rectum and the *vas deferens* (Fig. 8.7C). The gland cells produce refractive granules that may entirely fill the ducts.

References

ABOUL-EID, H.Z. (1969). Histological anatomy of the excretory and reproductive systems of *Longidorus macrosoma*. *Nematologica* 15, 437-450. DOI: 10.1163/187529269X00777

ANDRÁSSY, I. (1986). The genus *Mesodorylaimus* Andrássy, 1959 and its relatives (Nematoda: Dorylaimidae). *Acta Zoologica Academiae Scientiarum Hungaricae* 33, 207-261.

ANDRÁSSY, I. (1997). Nematodes in the Hévízi Lake (Hungary). *Állattani Közlemények* 82, 13-27.

BACCETI, B., DALLAI, R., GRIMALDI DE ZIO, S. & MARINARI, A. (1983). The evolution of the nematode spermatozoon. *Gamete Research* 8, 309-323.

BAJAJ, H.K. & JAIRAJPURI, M.S. (1978). Males of *Xiphinema opisthohysterum* Siddiqi, 1961 and *X. brevicolle* Lordello & Da Costa, 1961 with abnormal gonads. *Nematologica* 24, 334-337. DOI: 10.1163/187529278X00335

COOMANS, A. & LOOF, P.A.A. (1986). Observations on the glands of the male reproductive system in dorylaims and its phylogenetic importance. *Revue de Nématologie* 9, 261-265.

DAS, V.M. (1963). Morphological studies of the male *Nygolaimus aquaticus* Thorne, 1930. *Canadian Journal of Zoology* 41, 725-732.

JAIRAJPURI, M.S. & AHMAD, W. (1992). *Dorylaimida. Free-living, predaceous and plant-parasitic nematodes*. New Delhi, India, Oxford & IBH Publishing Co. Pvt. Ltd.

NAIR, P. (1973). The genus *Axonchium* (Nematoda: Belondiridae). III. Species with vaginal sclerotization (part 1). *Nematologica* 19, 169-189. DOI: 10.1163/187529273X00303

NAIR, P. (1975). The genus *Axonchium*. V. Species with amalgamated lip region. *Nematologica* 20(1974), 375-394. DOI: 10.1163/187529274X00014

NAIR, P. & COOMANS, A. (1973). The genus *Axonchium* (Nematoda: Belondiridae). I. Morphology, juvenile stages, diagnosis and list of species. *Nematologica* 18(1972), 494-513. DOI: 10.1163/187529272X00106

NAIR, P. & COOMANS, A. (1974). The genus *Axonchium* (Nematoda: Belondiridae). IV. Species with vaginal sclerotization (part 2). *Nematologica* 20, 69-87. DOI: 10.1163/187529274X00609

PEÑA-SANTIAGO, R., ABOLAFIA, J. & ÁLVAREZ-ORTEGA, S. (2014). New proposal for a detailed description of the dorylaim spicule (Nematoda: Dorylaimida). *Nematology* 16, 1091-1095. DOI: 10.1163/15685411-00002834

YUSHIN, V.V. (2010). Spermatid and spermatozoon structure in fresh-water nematode *Paractinolaimus microdentatus* (Nematoda: Dorylaimida: Actinolaimidae). *Russian Journal of Nematology* 18, 199-208.

Nematology Monographs & Perspectives, 2021, Vol. 13, 123-135

Chapter 9

Nervous system and receptors

The nervous system of dorylaims is made of a complex of delicate structures that are difficult to observe and have received little attention. Its general scheme was, however, beautifully described and illustrated by Anderson (1966; Fig. 9.1) in *Aporcelaimus amphidysis* (subsequently regarded as a synonym of *A. eurydoris*), whereas its ultrastructure was studied in detail by Roggen *et al.* (1967) in *Xiphinema index* and by Lippens *et al.* (1974) in *Aporcelaimellus obtusicaudatus*. The text in this chapter is therefore based mainly on, and/or taken from, these three contributions[1].

Central nervous system

Following Anderson's (1966) ideas, the central nervous system consists of nerve ring(s) and other associated elements. The nerve ring is a bundle[2] of nerves located at the anterior body region. In dorylaims, it encircles the slender (anterior) section of the pharynx, which becomes a little narrower at this level and is often at about one-third of the total neck length from the anterior end. A set of ganglia and commissures are associated with the nerve ring, but they are very difficult to observe and have only occasionally been reported in the literature, so that the available information is very often fragmentary and not always in agreement. Thus, Heyns (1964), when describing the new genus and species *Vanderlindia duplopapillata*, mentioned (and illustrated) the existence of three large (about 10 μm in diameter) ganglia, one dorsomedian and two subdorsal, located just posterior to the nerve ring,

[1] Lippens *et al.* (1974) pointed out that their scheme, based on TEM study, differed from that proposed by Roggen *et al.* (1967) and Anderson (1966), the latter based on LM observations, in some relevant aspects.

[2] Exceptionally, Goodey & Hooper (1963) recorded the existence of two nerve rings in a few members of the family Longidoridae.

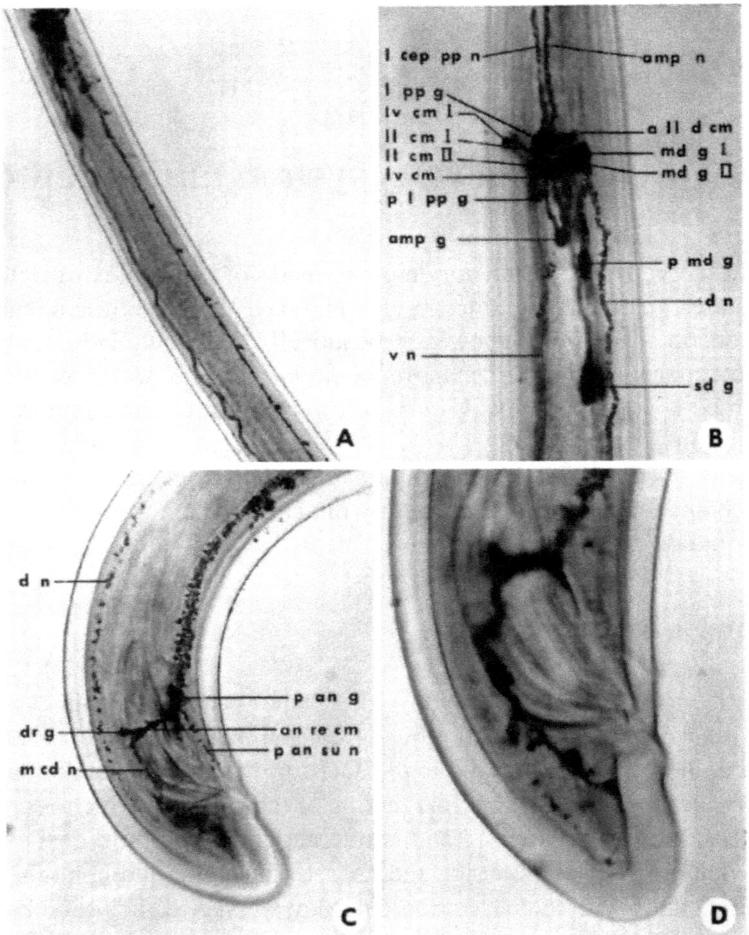

as well two smaller subdorsal ganglia directly anterior to the nerve ring, all being surrounded by a multicellular organ made up of nervous tissue. A ventro-lateral commissure at the level of the nerve ring, termed the hemizonid, and especially obvious in other nematode orders (for example tylenchs and rhabdits), has been mentioned by some authors (Goodey & Hooper, 1963; Siddiqi, 1970) in several dorylaims; it is perceptible due to the presence of a clear area in the hypodermis on the ventral side of body. A short distance posterior to the hemizonid a similar commissure may occur, the hemizonion (Goodey, 1963).

Anderson (1966) provided a very detailed description of these elements (Figs 9.1B; 9.2), but it remains uncertain if it is valid for dorylaims in general. Associated with the nerve ring are commissures and ganglia that appear to originate from paired lateral ganglia: the major and minor internolateral ganglia. The major internolateral ganglia lie immediately posterior to the nerve ring and are connected by two (I & II) laterolateral commissures from which arise the dorsal nerve and three ganglia. A pair of subdorsal ganglia, which lie near the level of the pharyngeal enlargement, also originate from the laterodorsal commissure II. A large post-middorsal ganglion, located interior to the dorsal nerve, originates from the middorsal ganglion I of laterolateral commissure I. Ventrally, the major internolateral ganglia are joined by another laterolateral commissure. The minor internolateral ganglia lie anterior to the nerve ring and are joined by dorsad and ventrad laterolateral commissures, which in turn are joined to laterolateral commissure I by way of paired oblique laterosubdorsal commissures. Lateroventral commisures I, connecting the ventral nerves with the amphidial ganglia (see below), emerge from the ventral chord at about the level of the nerve ring, separate right and left, and re-enter the hypodermis at the lateral hypodermal chord anterior to the nerve ring from where they continue posteriorly and inwardly to join with the amphidial ganglia, which are located a short distance posterior to the nerve ring. Two (right and left) large papillary ganglia lie in front of the minor internolateral ganglia. Two other ganglia, the

Fig. 9.1. *Photomicrographs of nervous system of* Aporcelaimus amphydisis. *A: Dorsal and ventral nerves; B: Laterodorsal view of central nervous system; C, D: Rectosympathetic nervous system of the male. (Abbreviations: a ll d cm = anterior laterolateral dorsal commissure, amp g = amphidial ganglion, amp n = amphidial nerve, an re cm = anorectal commissure, d n = dorsal nerve, dr g = dorsorectal ganglion, l cep pp n = lateral cephalic papillary nerve, l pp g = lateral papillary ganglion, ll cm I = laterolateral commissure I, ll cm II = laterolateral commissure II, lv cm = lateroventral commissure, lv cm I = lateroventral commissure I, m cd n = median caudal nerve, md g I = middorsal ganglion I, md g II = middorsal ganglion II, p an g = preanal ganglion, p an su n = preanal supplement nerve, p l pp g = postlateral papillary ganglion, p md g = postmiddorsal ganglion, sdg g = subdorsal ganglion; v n = ventral nerve.) (After Anderson (1966),* Canadian Journal of Zoology *44, pp. 818, 819, Fig. 2.)*

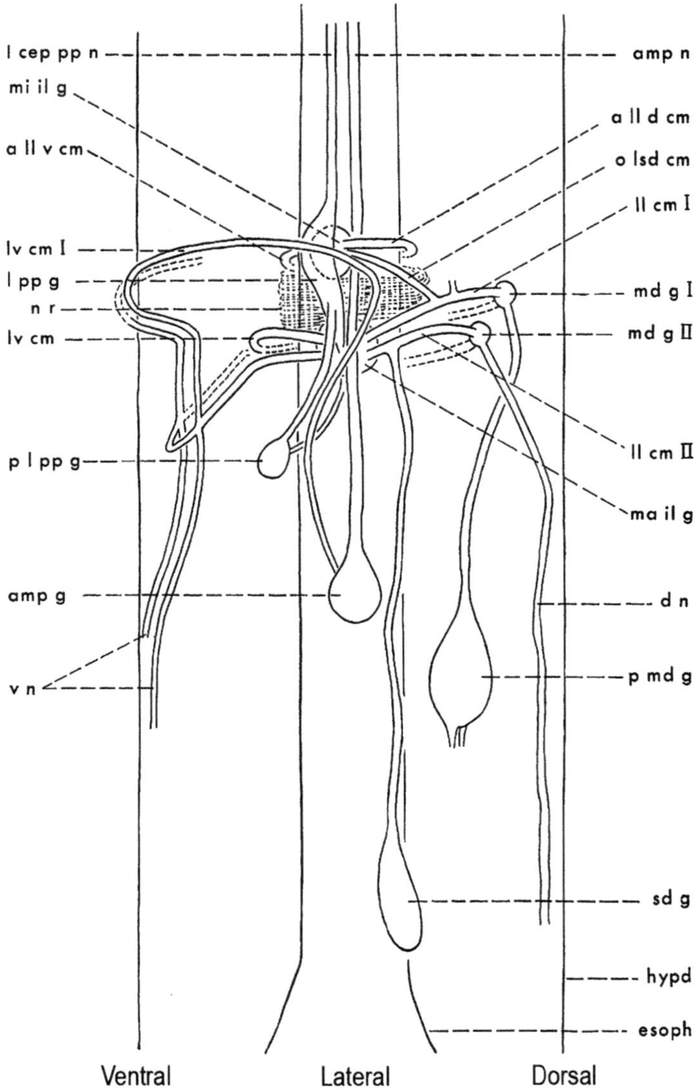

postlateral ganglia, extend from each papillary ganglion and these in turn connect with the major internolateral ganglia by way of a short commissure.

Nerves

According to Anderson (1966), the main nerves observed in *Aporcelaimus amphydisis* are the following. A large dorsal nerve originates at the middorsal ganglion II of lateral commissure II (Fig. 9.1A). From this ganglion, it runs posteriorly in the dorsal epidermal chord, ending near the tail tip (Fig. 9.1C). The paired ganglionated ventral nerves appear to originate at the major internolateral ganglia from where they run posteriorly in the ventral hypodermal chord (Fig. 9.1A). They pass to the right and left of the vagina and remain double to near their termination point at the preanal ganglion (Fig. 9.1C). Anteriorly, the ventral nerves connect with the amphidial ganglia by way of the lateroventral commissures I, thereby providing a direct line of communication with the amphids. One pair (right and left) of conspicuous amphidial nerves originate in their respective amphidial ganglion and end in the corresponding amphidial fusus at about the level of the odontostyle base. A large cephalic lateral papillary nerve lies to the ventral side of each amphidial nerve. Anteriorly, it terminates in each lateral lip, whereas posteriorly it ends in the corresponding lateral papillary ganglion.

Cephalic nervous system

The anterior region of dorylaims has a concentration of sense organs (labial and cephalic papillae, pores and amphids). The nervous

Fig. 9.2. *Semi-diagrammatic illustration of central nervous system of* Aporcelaimus amphydisis. *(Abbreviations: a ll d cm = anterior laterolateral dorsal commissure, a ll v cm = anterior laterolateral dorsal commissure, amp g = amphidial ganglion, amp n = amphidial nerve, d n = dorsal nerve, esoph = pharynx, hypd = hypodermis, l cep pp n = lateral cephalic papillary nerve, ll cm I = laterolateral commissure I, ll cm II = laterolateral commissure II, l pp g = lateral papillary ganglion, lv cm I = lateroventral commisure I, lv cm = lateroventral commisure, ma il g = major internolateral ganglion, mi il g = minor internolateral ganglion, md g I = middorsal ganglion I, md g II = middorsal ganglion II, n r = nerve ring, o lsd cm = oblique laterosubdorsal commissure, p l pp g = postlateral papillary ganglion, p md g = postmiddorsal ganglion, sd g = subdorsal ganglion, v n = ventral nerve.) (After Anderson (1966),* Canadian Journal of Zoology *44, pp. 818, 819, Fig. 3.)*

elements associated with these structures form the cephalic nervous system. According to Lippens *et al.* (1974), it mainly consists of eight nerve trunks: a dorsal, a ventral and two lateral, all of them embedded within the hypodermal chords, and four submedian trunks running through the pseudocoel. In addition, four intra-hypodermal nerves run in the submedian position posteriorly from the posterior insertion of the anterior (muscle) retractors, although their origin remains unknown.

The lateral nerve trunks (Fig. 9.3) comprise the axons and dendrites of the lateral internal and external labial nerves, amphidial nerves and the external lateral nerves, with accompanying glial elements. Several, irregularly spaced somatic receptors run from the lateral nerve trunk to the exterior. At a level corresponding to the base of the 'hydrostatic tissue' (*i.e.*, with the odontostyle retracted), a short nerve process containing a basal body, but without a cilium, is located on the outer margin of the trunk. From this level anteriad, the lateral nerve trunk divides. First, the external lateral nerve is detached, leading to three or four somatic receptors, ending in the most anterior of these receptors, which is located a short distance posterior to the amphidial aperture. Second, the amphidial nerve becomes separated from the complex of the papillary nerves. A short distance posterior to the nerve ring, the lateral labial nerves divide into an internal and an external branch that end in ducts that open to the exterior through pores in the centre of the inner and outer labial papillae.

The dorsal and ventral nerve trunks comprise the motor nerves and, in addition, the nerves that innervate the irregularly spaced somatic receptors. The nerve trunks end at the posterior edge of the dorsal and ventral arms of the cephalic framework.

The submedian nerve trunks (Fig. 9.3) originate from the submedian anterior ganglia. Where they cross the anterior retractors, the trunks are flattened, running radially from the pharyngeal wall and almost to the hypodermis of the body wall. Anteriorly, they once more become cylindrical. Somewhat posterior to the guiding ring, each of the trunks divides into a cephalic papillary nerve and an internal and external labial nerve, leading to the corresponding papillae.

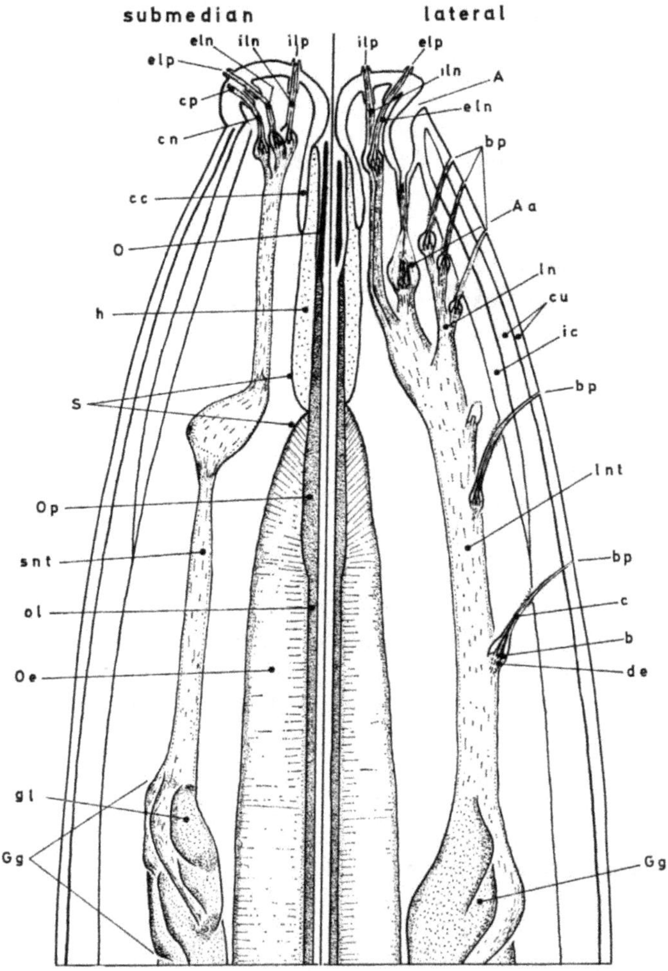

Fig. 9.3. *Reconstruction of the cephalic sensory system of* Aporcelaimellus obtusicaudatus. *(Abbreviations: A = amphid, Aa = amphidial ampulla, b = basal body, bp = body pore, c = sensory cilium, cc = cuticular cylinder, cn = cephalic nerve, cp = cephalic papilla, cu = cuticle, de = dendritic process, eln = external labial nerve, elp = external labial papilla, Gg = ganglion, gl = glial element, h = "hydrostatic tissue", ic = infracuticle, iln = internal labial nerve, ilp = internal labial papilla, ln = external lateral nerve, lnt = lateral nerve trunk, O = odontostyle, Oe = pharynx, ol = pharyngeal lining, Op = odontophore, S = stomodeal membrane, snt = submedian nerve trunk.) (After Lippens* et al. *(1974),* Nematologica *20, pp. 249, 256, Fig. 31.)*

Pharyngeal nervous system

A specialised nervous system is associated with the cuticular wall of the food canal, including the odontophore and the entire pharynx. The detailed organisation of this system was elucidated by Robertson (1979) in *Longidorus leptocephalus*. According to this author, there are three main nerves linking the odontophore with the pharyngeal basal bulb, and these nerves present several kinds of differentiations in the three main sections of their course, namely odontophore, slender portion of the pharynx, and basal bulb (Fig. 9.4). In the odontophore region the three main nerves divide in several branches with different destinations: four branches run anteriorly to form four cheilostom nerves associated with the odontostyle protractor muscles, another five also run anteriad, passing through the pseudocoel linking with the somatic nerve system, and one runs posteriad through the pesudocoel linking to the nerve ring. The region corresponding to the slender pharyngeal section contains three differentiations: an (anterior) sensory region, a soma of six neurons with prominent nuclei (in intermediate position), and a posterior commissure. In the basal bulb the three nerves divide to produce branches that innervate the pharyngeal glands (only three lead to their respective duct) as well the radial muscles of the bulb; the three nerves finally come together to form a small ring a short distance anterior to the pharyngeal base.

Rectosympathetic nervous system

The rectosympathetic nervous system consists of a few ganglia and commissures (*cf.* Anderson, 1966). In females, a pair of anorectal commissures originates from the preanal ganglion and extends dorsally right and left to the rectum, joining at the dorsorectal ganglion. A median caudal nerve extends into the tail from the dorsorectal ganglion. In the male, the cloacal commissures lie between the spicules and end dorsad to the cloaca in a pair of dorsorectal ganglia (Fig. 1C, D). The median caudal nerve extends ventrally into the tail and is ganglionated (Fig. 1D).

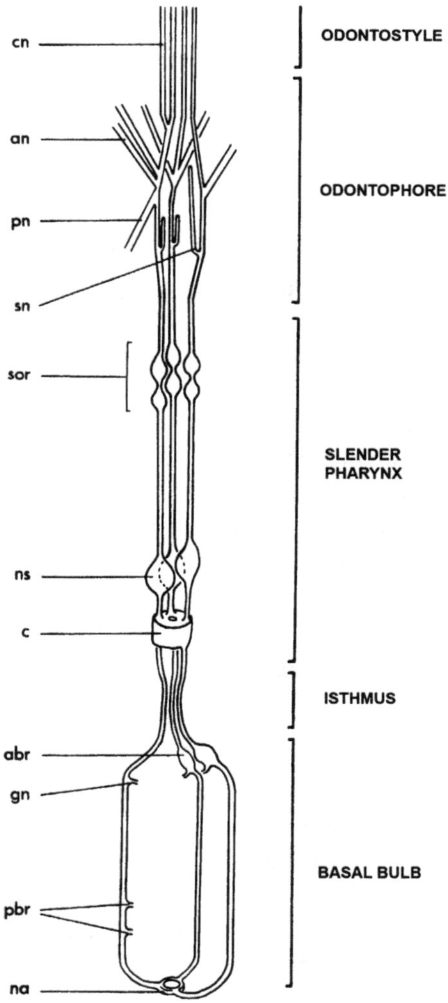

Fig. 9.4. *Schematic diagram showing pharyngeal nervous system of* Longidorus leptocephalus. *(Abbreviations: abr = anterior bulb receptor, an = nerve linking odontophore to hypodermal chord, c = commissure, cn = nerve to cheilostom, gn = dorsal gland nerve, na = nerve annulus around food canal, ns = nerve cell soma, pbr = posterior bulb receptor branches to cuticularised wall of food canal, pn = posterior nerve from odontophore to dorsal side of nerve ring, sn = ventral nerve in odontophore sinus, sor = slender pharyngeal sensory region.) (After Robertson (1979),* Nematologica *25, p. 250, Fig. 26.)*

Sensory structures

CHEMORECEPTORS

Amphids

These chemoreceptors[3] are characteristic of the nematode morphological pattern. As already known, and as seen with LM, each amphid consists of a fovea, amphidial canal and fusus. When observed with TEM (Lippens *et al.*, 1974), after their separation from the complex of the papillary nerves, the dendrites of each amphidial nerve divide in several nerve processes, on which the basal bodies of the sensory cilia are inserted. More than one cilium is inserted on some processes. The sensory cilia are enclosed in the amphidial ampulla (fusus) that leads anteriorly into the cuticularised amphidial duct. The amphidial duct forms two usually well-developed, but sometimes obscure, blind sac-like structures within the crests. In this area the ciliary processes have terminal filaments.

Other chemoreceptive elements

Chemoreceptive elements have been reported in the pharyngeal nervous system in both free-living forms and plant-parasitic species. Thus, Siddiqi (1971) observed the existence of special structures, termed endolids, in the subventral sectors of the cuticular lining of the slender pharyngeal section in several (free-living) species a short distance anterior to the nerve ring (Fig. 9.5), which appeared to have nerve terminals and were regarded as probable sensory organs. Similarly, Robertson (1976, 1979) noted the existence of ciliary bodies, regarded as chemoreceptors, whose nerve processes seemed to be associated with the odontophore and both pharyngeal sections in two longidorid species, and Bleve-Zacheo *et al.* (1993), who studied the fine structure of body pores, suggested that they represented chemosensilla and may serve to monitor the environment.

[3] Very often regarded as chemoreceptors, the amphids might play other roles as mechanoreceptors or secretory structures, as discussed by Coomans (1979).

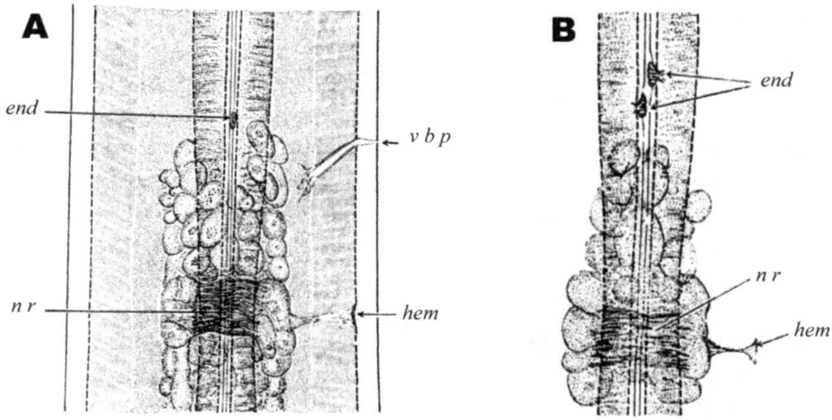

Fig. 9.5. *Nerve ring region of* Sicaguttur sartum *(A) and* Drepanodorylaimus *sp. (B), showing location of the endolids. (Abbreviations:* end = *endolids,* hem = *hemizonid,* n r = *nerve ring with associated ganglia,* v b p = *ventral body pore.) (After Siddiqi (1971),* Nematologica *16(1970), p. 489, Fig. 3C, D.)*

MECHANORECEPTORS

Labial and cephalic papillae[4]

As observed with LM, and especially with SEM, they are button-like structures, with a central pore surrounded by a variable number (few) of concentric incisures. TEM observations (Lippens *et al.*, 1974; Castanys-Cuello & López-Abella, 1977) show that each papilla consists of a nerve comprising four (labial papillary nerves) or three (cephalic papillary nerves) dendrites, surrounded by glial elements. The dendrites divide into ciliated processes that are enclosed within the ampullae. The fibrillar configuration is variable, usually irregular. The tip region of the cilia is modified to form terminal filaments, sometimes with two filaments per cilium. The ampulla leads into a membrane-line duct that becomes enclosed in a cuticular projection from the body wall. Between this cuticular outgrowth and the lining membrane a sheath of hypodermal tissue is present, entering the cuticle about half-way along the duct. In the region where the projections are located, the infracuticle and epidermis

[4] These sensory elements have traditionally been regarded as mechanoreceptors but, as explained in depth by Coomans (1979), they may have a bimodal (mechano- and chemoreceptive) function in many Dorylaimia.

intermingle in a very complicated way. The ducts open to the exterior through a pore.

According to Lippens *et al.* (1974), the somatic receptors are comparable structures to papillae and have a similar ultrastructural configuration. These receptors open in pores on the body surface and are arranged in one (sometimes two) rows throughout the lateral lines of the entire body, although dorsal and ventral pores may also be present, most often in the cervical region.

Other mechanoreceptive elements

Robertson (1979) suggested that the (chemo)sensory elements found in the pharyngeal nervous system might have a dual purpose, playing also a role (probably primary) as mechanoreceptors to detect the presence of food and/or monitor saliva flow from gland ducts, *etc.*

References

ANDERSON, R.V. (1966). Observations on the nervous system of *Aporcelaimus amphidysis* n. sp. (Nematoda: Dorylaimoidea). *Canadian Journal of Zoology* 44, 815-820.

BLEVE-ZACHEO, T., MELILLO, N.T., D'ADDABO, L. & ZACHEO, G. (1993). Fine structure of sense organs in the mid-region of *Xiphinema diversicaudatum* females. *Nematologica* 39, 496-504. DOI: 10.1163/187529293X00439

CASTANYS-CUELLO, F. & LÓPEZ-ABELLA, D. (1977). Órganos sensoriales de la regi6n anterior de *Mesodorylaimus bastiani* (Nematoda: Dorylaimida). *Nematologia Mediterranea* 5, 217-226.

COOMANS, A. (1979). The anterior sensilla of nematodes. *Revue de Nématologie* 2, 259-283.

GOODEY, J.B. (1963). *Soil and freshwater nematodes*, 2nd edition. London, UK, Methuen & Co.

GOODEY, J.B. & HOOPER, D.J. (1963). The nerve rings of *Longidorus* and *Xiphinema*. *Nematologica* 9, 303-304. DOI: 10.1163/187529263X00539

HEYNS, J. (1964). *Vanderlindia duplopapillata* n. gen., n. sp. (Nematoda: Dorylaimoidea), with a note on certain organs in the oesophageal region. *Nematologica* 10, 301-305. DOI: 10.1163/187529264X00051

JAIRAJPURI, M.S. & AHMAD, W. (1992). *Dorylaimida. Free-living, predaceous and plant-parasitic nematodes*. New Delhi, India, Oxford & IBH Publishing Co. Pvt. Ltd.

LIPPENS, P.L., COOMANS, A., DE GRISSE, A.T. & LAGASSE, A. (1974). Ultrastructure of the anterior body region in *Aporcelaimellus obtusicaudatus* and *A. obscurus*. *Nematologica* 20, 242-256. DOI: 10.1163/187529274X00212

ROBERTSON, W.M. (1976). A possible gustatory organ associated with the odontophore in *Longidorus leptocephalus* and *Xiphinema diversicaudatum*. *Nematologica* 21(1975), 443-448. DOI: 10.1163/187529275X00202

ROBERTSON, W.M. (1979). Observations on the oesophageal nerve system of *Longidorus leptocephalus*. *Nematologica* 25, 245-254. DOI: 10.1163/187529279X00262

ROGGEN, D.R., RASKI, D.J. & JONES, N.O. (1967). Further electron microscopic observation of *Xiphinema index*. *Nematologica* 13, 1-16. DOI: 10.1163/187529267X00850

SIDDIQI, M.R. (1971). *Oriverutus lobatus* gen. n., sp. n. and *Sicaguttur sartum* gen. n., sp. n. (Nematoda: Dorylaimoidea) from cultivated soils in Africa. *Nematologica* 16(1970), 483-491. DOI: 10.1163/187529270X00667

Nematology Monographs & Perspectives, 2021, Vol. 13, 137-146

Chapter 10

Caudal region (tail)

General concept

One of the diagnostic traits of the nematode body plan is the presence of a caudal region, also termed the tail, extending from the anus (in females) or cloacal aperture (in males) to the posterior extremity. In dorylaims, the caudal region is extraordinarily variable (Fig. 10.1) in both shape and size (length) and is a very useful feature for separating taxa. In addition, it may show significant differences among sexes (sexual dimorphism) and/or display important changes in shape during postnatal development.

Tail shape

There are three basic kinds of caudal region depending on overall shape, *i.e.*, filiform, conical and rounded.

Filiform tails (Fig. 10.2) are long, very often more than five times the body diameter at the level of the anus/cloacal aperture, and gradually taper throughout their length (Fig. 10.2A, C, D) before ending in a finely rounded (seldom acute) tip. This general pattern shows much variation. Being gradual, the tapering of tail may be more obvious in its anterior part (about one body diameter posterior to the anus/cloacal aperture) and much more gradual along the rest of its length (Fig. 10.2B, E, F). The profile of the tail may be nearly straight (Fig. 10.2A, B, F), but it usually appears more or less regularly curved ventrad somewhere along its total length, either in the anterior or posterior section (Fig. 10.2C), or only at the terminus (Fig. 10.2A, D); more exceptionally, it may be curved dorsad (Fig. 10.2E). Other very special differentiations occasionally occur in long tails. For example, in some species, the tip appears visibly thickened, becoming wider than the adjacent part of the tail, thereby forming the so-called filiform, clavate tail (Fig. 10.2F). In some representatives of the genera *Falciasta* and *Oxydirus* the filiform

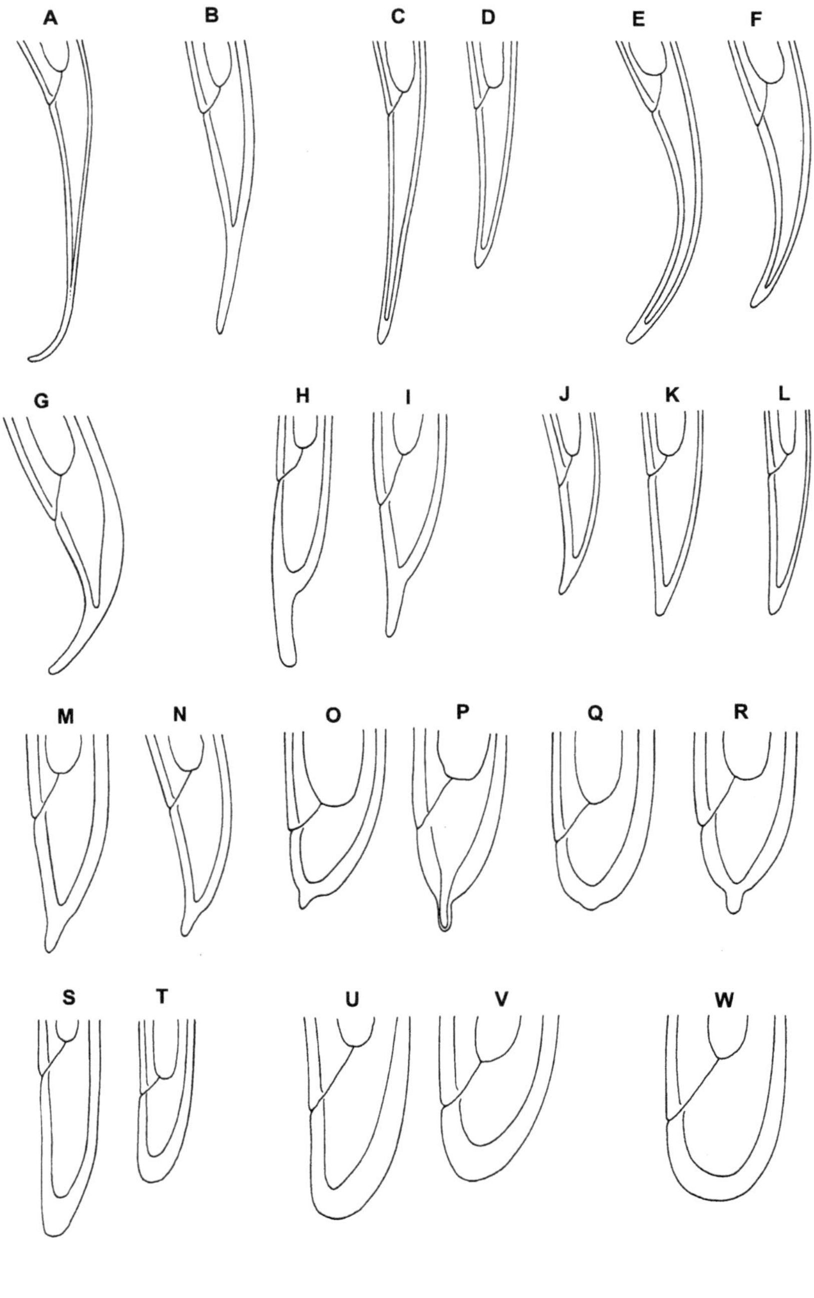
A
B
C
D
E
F
G
H
I
J
K
L
M
N
O
P
Q
R
S
T
U
V
W

tail exceptionally bears visible, 'wing-like' thickenings of the cuticle located at about the middle region of the tail (Fig. 10.2C). In general, the inner core of the tail does not extend along its entire length, a more or less developed hyaline portion, which is devoid of visible musculature and other components of body wall, occupying the terminal region (Fig. 10.2B, C, F).

The conical tails (Fig. 10.3) are comparatively short (very often less than three anal body diameters), tapering more or less gradually to a finely rounded (Fig. 10.3A, D, F) or acute (Fig. 10.3B, G, H) end, and regularly curved ventrad, that is, concave along its ventral contour and convex on its dorsal contour. Nevertheless, and leaving aside its size (length), the conical tail shows some interesting variants, mainly concerning its profile. Its curvature ranges from weak to strong, nearly hook-like (Fig. 10.3A-C). Sometimes it does not display any appreciable curvature, its ventral side appearing nearly straight and the dorsal slightly convex (Fig. 10.3D, E). Less frequently, it becomes more or less re-curved, that is, bent dorsad in its posterior part, with the ventral contour appearing straight or weakly convex whereas the dorsal contour becomes convex in its anterior part and concave in its posterior region (Fig. 10.3F-I). The inner core of the tail may reach the tip (Fig. 10.3E, F) but a more or less developed hyaline portion is very often distinguishable (Fig. 10.3A, B, D, G).

Many dorylaimid taxa bear a rounded tail. It is short (very usually up to one anal/cloacal body diameter long), convex on both sides, and broadly blunt at the terminus (Fig. 10.4). Nevertheless, some variation is also found in this shape, which ranges from slightly longer (rounded conoid, Fig. 10.4A-C) to very short (nearly hemispherical, Fig. 10.4D, E). Sometimes, the rounded tail may present a short, terminal, finger-like, projection, the so-called digitate tail (Fig. 10.1O-R). Occasionally, the tail is much longer than the anal/cloacal body diameter and with

Fig. 10.1. *Different types of tail shape as found in juveniles and adults in species of the genus* Xiphinema *(after Coomans* et al. *(2001).* Annales Sciences Zoologiques, Musée Royal de l'Afrique Centrale, Tervuren, Belgique, *287: p. 31, Fig. 11). A: Filiform; B-D: Conical elongate, ventrally straight; E-G: Conical elongate, curved ventrad; H, I: Conical elongate with a dorsal concavity; J-N: Conical; O-R: Short and rounded with a subdigitate (peg-like) terminus; S, T: Subcylindrical; U, V: Short and convex conoid; W: Short and rounded.*

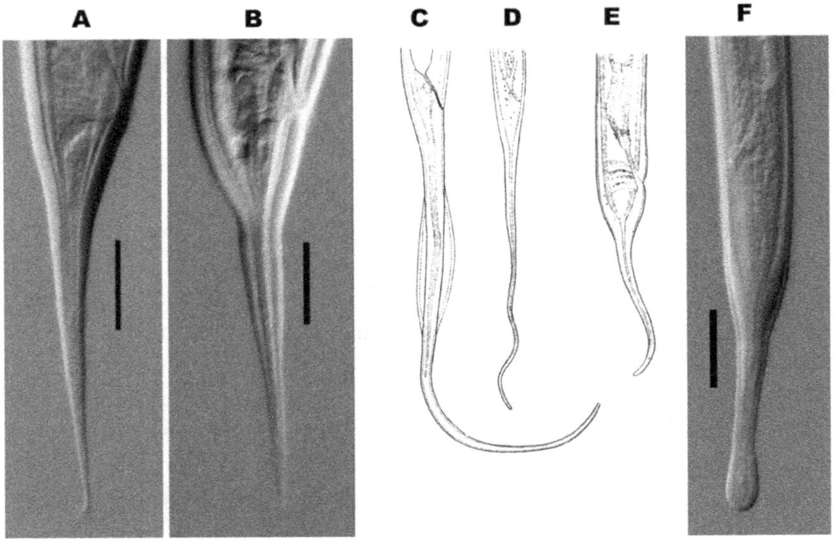

Fig. 10.2. *Filiform tails in dorylaims. A:* Dorylaimus *sp. from Spain; B:* Mesodorylaimus *sp. from Spain; C:* Falcihasta palustris*; D:* Timmus tenerhatus*; E:* Thornenema caudatum*; F:* Belondira *sp. from Vietnam. (C-E: After Jairajpuri & Ahmad (1992),* Dorylaimida*, p. 51, Fig. 23A, B, D.) (Scale bars: A = 50 μm; B-F = 20 μm.)*

the terminus broadly rounded, the so-called cylindrical tail. More exceptionally still, either the whole tail or only its posterior portion (in cylindrical tails) can become visibly broader (more swollen) than the adjacent part of the body/tail to form the clavate tail (Fig. 10.4F).

The existence of some kind of malformation in tail shape is not unusual. The most common case is a shortening of long tails resulting in a kind of stump. Much more exceptional is tail splitting as observed by Dhanam *et al.* (1994) in a juvenile of *Paraoxydirus* sp.

Tail sexual dimorphism

In dorylaims, the morphology of the caudal region very often is similar in both (adult) sexes, with only minor differences affecting its size (length), shape and ventrad curvature, the latter usually more appreciable in males. This scheme represents the general pattern of the group. A few genera, however, are characterised by having a marked sexual dimorphism in tail shape, *e.g.*, long and filiform in the female,

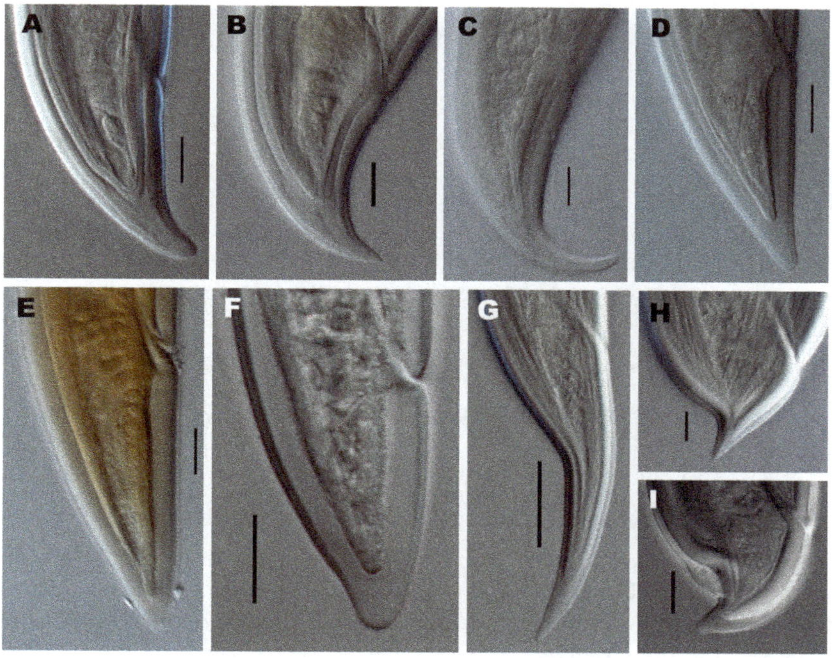

Fig. 10.3. *Conical tails in dorylaims. A:* Heterodorus transilvanicus *from Romania; B:* Rhysocolpus iuventutis *from Spain; C:* Eudorylaimus *sp. from Spain; D:* Dorydorella bryophila *from the USA; E:* Metaporcelaimus insignis *from Sri Lanka; F:* Ecumenicus monohystera *from Spain; G:* Aporcelinus elongicaudatus *from Costa Rica. H:* Aporcelinus amazonicus *from Costa Rica. I:* Aporcelinus falcicaudatus *from Vietnam. (Scale bars = 10 μm.)*

short and rounded in the male (Fig. 10.5A, B). This alternative pattern has been reported in some members of at least five dorylaimid families, namely Actinolaimidae, Belondiridae, Dorylaimidae, Mydonomidae and Thornenematidae, and it is certainly the expression of a convergence process occurring in several taxa and/or at different times in the evolutionary history of dorylaims, but whose adaptive function remains obscure. Remarkably, male juvenile stages of these taxa bear long tails (see below). The genus *Afrodorylaimus* is the only one presenting a rather peculiar tail sexual dimorphism, having long-tailed females and conical-tailed (never rounded-tailed) males (Fig. 10.5C, D).

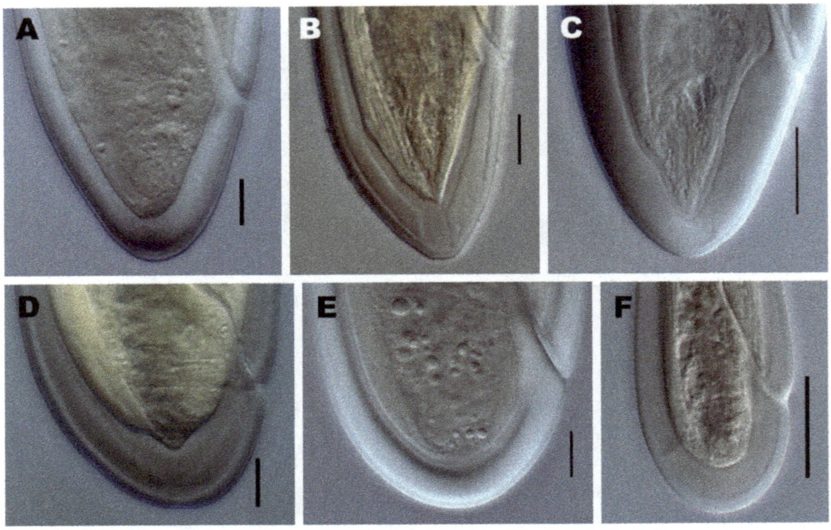

Fig. 10.4. *Rounded tails in dorylaims. A:* Aporcelaimellus rotundus *from Spain; B:* Enchodelus poronensis *from Romania; C:* Nevadanema nevadense *from Spain; D:* Aporcelaimellus brevicaudatus *from Spain; E:* Labronema neopacificum *from Vietnam; F:* Crassolabium rhopalocercum *from Vietnam. (Scale bars: A, B, D, E = 10 μm; C, F = 20 μm.)*

Postembryonic changes in tail shape

Unfortunately, the study of postembryonic (or postnatal) development in dorylaims has not received enough attention. Nevertheless, available information shows some interesting developmental patterns in which significant changes in the morphology of the caudal region take place. In the most common pattern, the shape of the caudal region of the four juvenile stages does not differ significantly in shape from that of their adults. There are, however, at least two other alternative patterns: one, observed in some long-tailed members of Dorylaiminae with sexual dimorphism, and another reported in rounded-tailed members of Labronematinae. Similar or comparable developmental patterns have been also reported in other dorylaimid taxa, such as the genus *Xiphinema* in Longidoridae (Coomans *et al.*, 2001) and *Enchodelus babakicus* in Nordiidae (Pedram *et al.*, 2009).

Shafqat *et al.* (1991) explained and illustrated the post-embryonic development of *Dorylaimus stagnalis* (Fig. 10.6A), an amphimictic

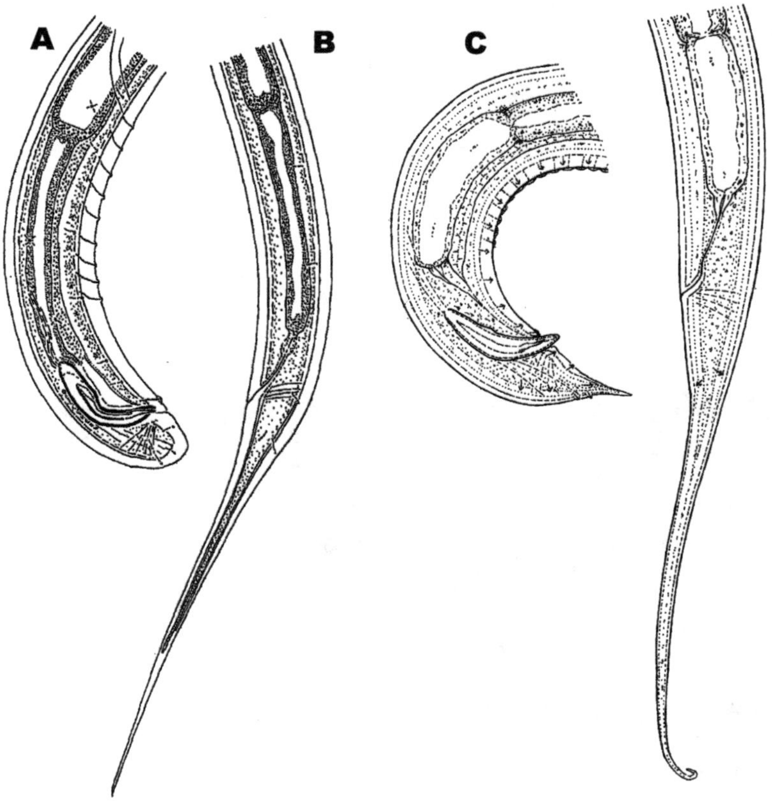

Fig. 10.5. *Sexual dimorphism of tail shape in dorylaims. A, B: Posterior body region of male (A) and female (B) in* Neoactinolaimus proximus *(after Vinciguerra & Coomans (1991),* Revue de Nématologie *14, p. 592, Fig. 2E, F); C, D: Posterior body region of male (C) and female (D) in* Afrodorylaimus bwana *(after Andrássy (2012),* Journal of Nematode Morphology and Systematics *5, pp. 166, 167; Figs 7D, 8C).*

species with sexual dimorphism. The four juvenile stages display a long filiform tail in both the female and the male. After the fourth (last) moult, the long tail of the male fourth-stage juvenile (J4) becomes short and rounded in the adult male, involving an abrupt change in the general morphology of the nematode. A short time before the last moult, when the adult cuticle is produced, the new rounded tail of the developing adult male is already easily visible inside the caudal region of the J4.

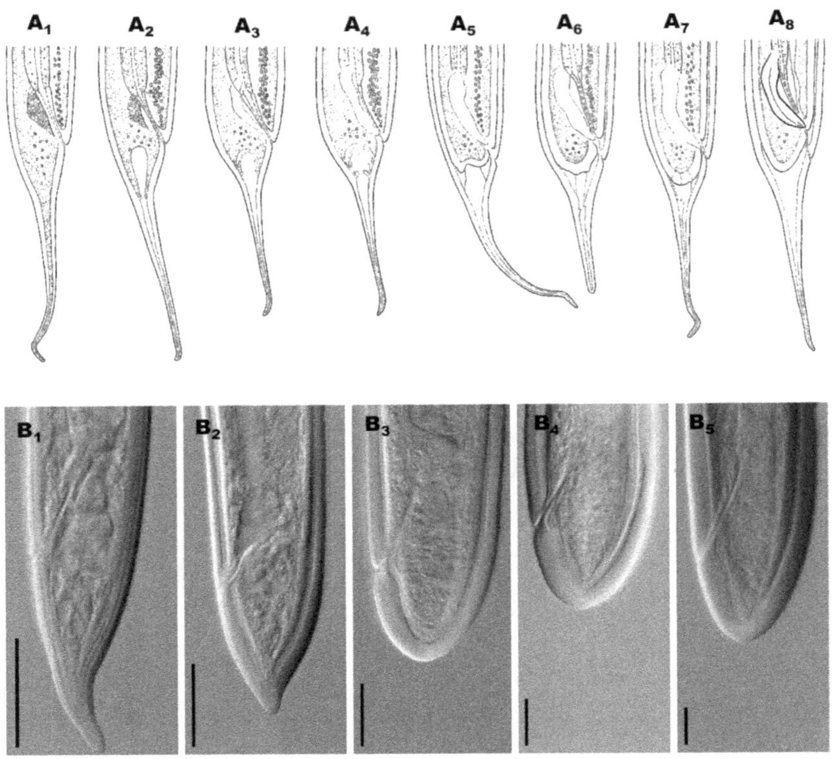

Fig. 10.6. *Postnatal development of tail in dorylaims. A: Posterior region of third- (A_1), early fourth- (A_2), midfourth- (A_3, A_4), late fourth- (A_5), moulting fourth- (A_6, A_7) and fourth-stage (just before final moulting, A_8) male juvenile stages in* Dorylaimus stagnalis *(after Shafqat* et al. *(1991),* Revue de Nématologie *14, p. 69, Fig. 6); B: Caudal region of first- (B_1), second- (B_2), third- (B_3), fourth-stage (B_4) juveniles and adult (B_5) in* Nevadanema nevadense *(after Álvarez-Ortega & Peña-Santiago (2013),* Journal of Nematode Morphology and Systematics *16, p. 169, Fig. 1F-J.) (Scale bars = 10 μm.)*

The post-embryonic development of rounded-tailed taxa also shows remarkable changes affecting the shape of the caudal region. One or more juvenile stages of both sexes bears a long or elongate tail, which, after a more or less progressive shortening process, results in no, one, or more than one, rounded-tailed juvenile and, invariably, in the rounded-tailed adult (Ferris, 1968; Álvarez-Ortega & Peña-Santiago, 2013).

Two other remarkable aspects concerning the postnatal development of the caudal region should be pointed out. First, there is not a single documented case of tail shape reversal (from rounded-tailed forms to long-tailed forms) during postembryonic development. Thus, it may be hypothesised that once the long tail shape has disappeared, it does not reappear. Second, conical-tailed forms do not undergo any relevant change in their caudal region during postnatal development, although further studies should be conducted on this topic.

Functional and evolutionary aspects of tail shape

Coomans *et al.* (2001) discussed many aspects of the nature of the tail in the genus *Xiphinema*, but their arguments and subsequent discussion are also valid for dorylaims in general. A long tail is assumed to be advantageous for swimming and characteristic of those forms dwelling in aquatic or wet habitats. The evolution of short tails was probably caused by the transition to more terrestrial habitats, a process that, as mentioned above, occurred several times during the evolutionary history of Dorylaimida. Nevertheless, a mixture of long- and short-tailed taxa is very often found in any kind of habitat. This means that not only the tail shape, but other ecological and/or functional aspects of dorylaimid biology should be explored and tested to explain such community assemblages.

The hypothesis that long tails represent a primitive (plesiomorphic) state is corroborated by both embryological and ecological evidences. On the one hand, as demonstrated in *Xiphinema*, and in some other dorylaims (see above), long tails always precede short tails when both types occur during ontogeny, with no documented reversal process. On the other hand, long-tailed dorylaimid forms appear to be predominantly freshwater denizens, whereas short-tailed forms are mainly terrestrial.

References

ÁLVAREZ-ORTEGA, S. & PEÑA-SANTIAGO, R. (2013). The postembryonic development of *Nevadanema nevadense* (Dorylaimida, Qudsianematidae) reveals more relevant changes than expected. *Journal of Nematode Morphology and Systematics* 16, 167-172.

ANDRÁSSY, I. (2012). The genus *Afrodorylaimus* Andrássy, 1964 and its species (Dorylaimida: Dorylaimidae: Afrodorylaiminae). *Journal of Nematode Morphology and Systematics* 15, 149-185.

COOMANS, A., HUYS, R., HEYNS, J. & LUC, M. (2001). Character analysis, phylogeny and biogeography of the genus *Xiphinema* Cobb, 1913 (Nematoda: Longidoridae). *Annales de Sciences Zoologiques, Musée Royal de l'Afrique Centrale, Tervuren, Belgique* 287, 1-239.

DHANAM, M., JAIRAJPURI, M.S. & KUMAR, A.C. (1994). Unusual tail splitting in a juvenile of *Paraoxydirus* sp. *Afro-Asian Journal of Nematology* 4, 238-239.

FERRIS, V.R. (1968). Biometric analyses in the genus *Labronema* (Nematoda: Dorylaimida) with a description of *L. thornei* n. sp. *Nematologica* 14, 276-284. DOI: 10.1163/187529268X00480

JAIRAJPURI, M.S. & AHMAD, W. (1992). *Dorylaimida. Free-living, predaceous and plant-parasitic nematodes*. New Delhi, India, Oxford & IBH Publishing Co. Pvt. Ltd.

PEDRAM, M., NIKNAM, G., GUERRERO, P., YE, W. & ROBBINS, R.T. (2009). Morphological and molecular characterisation of *Enchodelus babakicus* n. sp. and *E. macrodorus* Thorne, 1939 (Nematoda: Nordiidae) from Iran. *Nematology* 11, 895-907. DOI: 10.1163/156854109X430563

SHAFQAT, S., JAIRAJPURI, M.S. & BILGRAMI, A.L. (1991). Developmental biology of *Dorylaimus stagnalis* Dujardin, 1845 (Nematoda: Dorylaimida). *Revue de Nématologie* 14, 61-71.

VINCIGUERRA, M.T. & COOMANS, A. (1991). *Hexactinolaimus* Yeates, 1973, a junior synonym of *Paractinolaimus* Meyl, 1957 and a revised classification of Actinolaimidae (Nematoda: Dorylaimida). *Revue de Nématologie* 14, 589-596.

Nematology Monographs & Perspectives, 2021, Vol. 13, 147-169

Chapter 11

Feeding habits and feeding behaviour

Methodological constraints

Feeding is one of the lesser known aspects of dorylaimid biology. Available information derives from different kinds of experimental and/or observational studies.

Direct observations on the feeding behaviour of live specimens, either extracted from soil samples or maintained *in vitro*, occurred very early in the history of Nematology. Thus, de Man (1884) reported a *Dorylaimus* specimen piercing the cuticle of a *Cephalobus* individual.

Much early evidence, however, was based on the examination of gut contents in fixed and mounted specimens, where remains of food resources, for example, oligochaete setae or green materials, were easily recognisable within the intestine of the nematodes. Nielsen (1949) mentioned that the food of some *Dorylaimus* species consisted of plant juices, which they obtained by piercing vegetable tissues and sucking out their contents as the gut was often a green colour from the chlorophyll.

In spite of the technical difficulties involved in culturing dorylaims *in vitro*[1], their maintenance in Petri dishes and other media for a period of time produced an extraordinary advance in the study of their feeding habits. Nielsen (1949) was probably the first author able to obtain positive results culturing several dorylaimid species, *Aporcelaimellus obtusicaudatus* and *Eudorylaimus carteri* among them, in water with different substrates (algae, bacteria, moss leaves). Culturing on agar plates thus facilitated many new observations.

Feeding experiments were designed to analyse some basic aspects of the feeding biology of dorylaims, including the spectrum of food sources, food preference, feeding frequency, *etc.* (Nielsen, 1949; Esser,

[1] As dorylaims are recognised as K-strategists (comparatively large nematodes with long life cycles and low reproduction rates), they are very sensitive to any change in their habitat and, *a priori*, they are not good candidates for culturing.

1963; Wood, 1973, 1974; Grootaert & Small, 1982; Ferris & Ferris, 1989; Okada *et al.*, 2006).

Feeding behaviour of particular species were later studied, especially in predatory taxa, resulting in the elucidation of interesting issues such as the feeding action procedure, the role played by the stomatal and pharyngeal structures, *etc.* (Linford & Oliveira, 1937; Grootaert & Wyss, 1978; Khan *et al.*, 1991; Bilgrami, 2008).

More recently, PCR molecular techniques have been used to determine the feeding spectrum of a few predatory dorylaims by identifying the nature of their gut contents (Cabos *et al.*, 2013).

A review of the available literature reveals that the current information concerning the feeding habits of dorylaims is rather limited for several reasons. First, only 61 (out of more than 3000 valid) species, belonging to 38 (out of 289 valid) genera, have had some type of study on this topic (Tables 11.1, 11.2), which represents a very low percentage. Second, data appear unbalanced in favour of presumably predacious taxa (Table 11.1) as their number is significantly higher than those obtained for non-predacious forms (Table 11.2). Third, a few genera (*Aporcelaimellus*, *Labronema* and *Tylencholaimus* among others) have become over-represented, whilst many others are absent from the list or show few or nearly anecdotal references. In spite of these restrictions, it is possible to extract some interesting ideas that show the versatility of dorylaims in relation to feeding habits.

Feeding spectrum

Leaving aside the general assumption that dorylaims are predominantly omnivorous or predacious nematodes (for example, see the compilation by Yeates *et al.*, 1993), an idea mainly based on classical references – Thorne (1939) is an inestimable source of basic data -, experimental studies have demonstrated and/or corroborated the fact that the feeding spectrum of the group is more diverse than expected. It is important to say that, as stated by Yeates (1971: p. 178), "*in vivo* and *in vitro* (observations) are complementary and rarely contradictory. However, *in vitro* results do not prove that a particular species obtains all, or even some, of its nourishment in this way in the soil". However, there is enough evidence to assert with confidence that the feeding habits of dorylaims show a wide range, from predation to microherbivory.

Table 11.1. *A compilation of dorylaimid predatory species and their corresponding prey, with indication of the type of study carried out and the respective bibliographic references.*

Predatory family and species	Prey species or prey remains[1]	Study type[2]	Original reference[3]	Later reference or compilation[3]
Actinolaimidae				
Actinolaimus spp.	*Melm*, nematode eggs, rotifers	C	1, 2	3, 4, 5, 6, 7, 8
	Pan	C	9	
Brasilaimus bryophilus	Dead mite	D	10	8
Neoactinolaimus agilis	*Ant, Aphs, Hei, Hem, Hio, Hoi, Meli, Parc, Pars, Roro, Tylm, Tyls, Xia*	B, C, E	11	12
Neoactinolaimus duplicidentatus	*Glr, Meln, Pan*	C, E	13	12
Neoactinolaimus sp.	*Meli, Ras, Rore*	M	14	
Paractinolaimus elongatus	*Ant, Hio, Hoi, Los, Meli, Tylm, Xib*	C, E	15	12
Paractinolaimus vigor	*Ces*	C	10	8
Westindicus rapax	*Xis*, dead enchytraeids, dead tardigrade	D	10	8, 12
Aporcelaimidae				
Aporcelaimellus amylovorus	*Tyls*	T	16	8
Aporcelaimellus nivalis	*Hio, Rhs*	B, C	17	12
	Acs, Acsp, Ant, Aphs, Apn, Aqt, Bas, Ces, Chs, Dos, Hei, Hem, Hemd, Hetm, Hio, Hoi, Los, Meli, Mesb, Monf, Monl, Mono, Mym, Pan, Parc, Rhs, Scs, Tylm, Xia, Xii	C, E	18	
	Hio, Rhs	B, C	19	
Aporcelaimellus obtusicaudatus[4]	*Hets* (eggs)	D	20, 21, 7	22, 8, 12
	Pan	C, D	23	
	Nematodes, mites, *Pan*	C	9	
Aporcelaimellus sp.	*Acsp, Apha*	C, E		8
	Meli, Ras, Rore	M	14	
Aporcelaimoides haguei	Oligochaete setae	G	10	8
Aporcelaimus regius	Oligochaete setae	G	24	25, 8
Aporcelaimus vorax	Oligochaetes	G	21	8
Aporcelaimus spp.	Small oligochaetes	G	21	3, 4, 6, 8
	Hett	D	26	8, 12
	Acsp, Apha	C	27	8
	Nematodes	C	28	8
	Bacteriophagous nematodes	D	7	12

Table 11.1. *(Continued.)*

Predatory family and species	Prey species or prey remains[1]	Study type[2]	Original reference[3]	Later reference or compilation[3]
Aporcelinus americanus[5]	*Ant, Apha, Aphs, Bas, Hetm, Hio, Los, Meli, Trs, Tylm, Tyls, Xib*	C, E	29	12, 30
	Acsp, Cys other nematodes, mites, enchytraeid forms	C, D, G	9	
Sectonema barbatoides	?	T	31	
Sectonema rotundicauda	?	T	4	8
Sectonema ventrale	Oligochaetes-Enchytraeids	D, G	25, 32	8
Sectonema sp.	Oligochaetes	D, G	25	3, 4, 8
	?	T	5	8
Belondiridae				
Axonchium sabulum	?	T	33	8
Dorylaimidae				
Crassolabium ettersbergense	Nematodes	C, D	58	
Crassolabium sp.	*Pan*	C	9	
Dorylaimus asymphydorus	Oligochates-tubificids	G	34	8
Dorylaimus stagnalis	*Acsp, Ces, Hei, Hio, Hoi, Rhs, Tylm*	B, C, E	35	12
	Acsp, Ant, Ces, Hei, Hem, Hemy, Hetm, Hio, Hoi, Los, Meli, Pan, Parc, Pars, Scs, Tylm, Xia	C, E	36	12
Dorylaimus spp.	Mite eggs	D	37	8
	Melm, nematode eggs, rotifers	C	1, 2	7, 8
	Ces[8]	T	38	39, 8
Halodorylaimus marinus	?Oligochaetes	D	21	8
	Pan	C, D	9	
Labronema ferox	Oligochaetes	G	32	8
	Rhs	C	40	8
	Nematodes	C	2, 41	42, 8
Labronema thornei	*Rhs*	C	40	8
Labronema vulvapapillatum	*Ant, Apha, Pan,*	C	43	8, 12
	Pan	C	44	8, 12
	Glr, Meln, Pan	C, E	13	7, 8, 12
	Mono (eggs), mites, copepod (nauplius larva)	C	8	

Table 11.1. *(Continued.)*

Predatory family and species	Prey species or prey remains[1]	Study type[2]	Original reference[3]	Later reference or compilation[3]
Labronema spp.	Nematodes, oligochaetes, oligochaete eggs, mite, rotifer egg	C	45	6, 8
	?	G	5	8
	Acsp, *Apha*	C, E	27	8
	Hed, *Hemt*, *Pan*	C, E	46	
	Pan		9	
Laimydorus baldus	*Hio*, *Meli*	B, C	47, 48	
Laimydorus serpentinus	Nematodes	D	21	8
Mesodorylaimus bastiani	*Acs*, *Acsp*, *Ant*, *Aphs*, *Apn*, *Aqt*, *Bas*, *Ces*, *Chs*, *Dos*, *Hei*, *Hem*, *Hemd*, *Hetm*, *Hio*, *Hoi*, *Los*, *Meli*, *Mesb*, *Monf*, *Monl*, *Mono*, *Mym*, *Pan*, *Parc*, *Pars*, *Rhs*, *Scs*, *Tylm*, *Xia*, *Xii*	C, E	49	12
	Tylm	B, C	50	
	Acsp, *Ant*, *Aphs*, *Aqt*, *Bag*, *Ces*, *Dim*, *Hei*, *Hem*, *Hetm*, *Hio*, *Hoi*, *Meli*, *Mesb*, *Rhs*, *Tylm*	B, C, E	51	
	Hio, *Rhs*	B, C	52, 53	
Mesodorylaimus cf. lissus	Nematodes	C, D	42	
Mesodorylaimus cf. pseudobastiani	Many nematode species, mites		9	
Mesodorylaimus spp.	Rotifers, ruptured collembolan, *Clarkus gerlachei* cadaver	D, G	54	8
	Tyls	T	8	
	Meli, *Ras*, *Rore*	M	14	
	Pan	C	9	
Nordiidae				
Pungentus engadinensis	*Tyls*	T	8	
Pungentus monohystera	*Tyls*		16	
Nygolaimidae				
Aquatides aquaticus	Oligochaetes-tubificid setae	G	25, 32	8

Table 11.1. *(Continued.)*

Predatory family and species	Prey species or prey remains[1]	Study type[2]	Original reference[3]	Later reference or compilation[3]
Aquatides thornei	*Chs, Rhs*	C	55	8, 12
	Acsp, Ant, Ces, Hei, Hem, Hemy, Hetm, Hio, Hoi, Los, Meli, Pan, Parc, Pars, Scs, Tylm, Xia	C, E	36	12
	Tylm	B, C	50	
	Acsp, Ant, Aphs, Aqt, Bag, Ces, Dim, Hei, Hem, Hetm, Hio, Hoi, Meli, Mesb, Rhs, Tylm	B, C, E	51	
	Hio, Rhs	B, C	53	
Feroxides ferox	Oligochaete setae and cuticle	G	25, 32, 56	8
Nygolaimus brachyuroides	?	T	31	8
Nygolaimus directus	Oligochaetes-enchytraeid setae	G	33	8
Nygolaimus sp.	Oligochaetes	G	25	3, 4, 6, 8
	Oligochaetes	D	32	8
	?	T	5	8
	?			57
	Nematodes	C	9	
Paravulvus sp.	*Acsp, Apha*	C, E	27	8
Solididens vulgaris	Oligochaetes-enchytraeids	D, G	25	8
Qudsianematidae				
Carcharolaimus dentatus	Small oligochaetes	?D, G	32	4, 8
Carcharolaimus formosus	Small oligochaetes	D		
Carcharolaimus sp.	*Acs*	C	45	6, 8
	Rore	D	7	
Caribenema pizai	Small oligochaetes	T	4	8
Discolaimium jairajpurii[6]	*Rhs*	D	32	8
Discolaimium sp.	Nematodes	T	3, 4	8
Discolaimus major[7]	*Hio, Meli*	B, C	33, 47, 48	8
Discolaimus silvicolus	*Ant, Apha, Aphs, Bas, Hetm, Hio, Los, Meli, Trs, Tylm, Tyls, Xib*	C, E	29	12
Discolaimus similis	*Dorylaimus* odontostyle	G	32	8
Discolaimus spp.	*Melm*, nematode eggs, rotifers	C, D	1, 2	7, 8

Table 11.1. *(Continued.)*

Predatory family and species	Prey species or prey remains[1]	Study type[2]	Original reference[3]	Later reference or compilation[3]
	Nematodes, oligochaetes, oligochaete eggs, mite, rotifer egg	C	45	3, 4, 7, 8, 6
	Tyls	G	8	
	Plectus sp.	C	9	
Eudorylaimus ? carteri	Mite eggs	D	37	42, 8
Eudorylaimus meridionalis	Many nematode species, mites	C	9	
Eudorylaimus spp.	*Acsp, Apha*	C, E	27	8
	Nematodes	T	59	8
	Nematodes	C	28	8
	Bux	D	7	
Thorniidae				
Thornia sp.	*Apha, Tyls* (juv.), nematodes	C	59	7, 8, 12

[1] Abbreviations used to identify the nematode preys in each observation or experiment: *Acs* = *Acrobeles* sp. *Acsp* = *Acrobeloides* sp. *Ant* = *Anguina tritici*. *Aphs* = *Aphelenchoides* sp. *Apha* = *Aphelenchus avenae*. *Apn* = *Aporcelaimellus nivalis*. *Aqt* = *Aquatides thornei*. *Bag* = *Basiria graminophila*. *Bas* = *Basiria* sp. *Bux* = *Bursaphelenchus xylophilus*. *Ces* = *Cephalobus* sp. *Chs* = *Chiloplacus symmetricus*. *Cys* = *Cylindrolaimus* sp. *Dim* = *Discolaimus major*. *Dos* = *Dorylaimus stagnalis*. *Glr* = *Globodera rostochiensis*. *Hed* = *Helicotylenchus dihystera*. *Hei* = *Helicotylenchus indicus*. *Hem* = *Hemicriconemoides mangiferae*. *Hemd* = *Hemicycliophora dhirenderi*. *Hemt* = *Hemicycliophora typical*. *Hemy* = *Hemicycliophora* sp. *Hetm* = *Heterodera mothi*. *Hets* = *Heterodera schachtii*. *Hett* = *Heterodera trifolii*. *Hio* = *Hirschmanniella oryzae*. *Hoi* = *Hoplolaimus indicus*. *Los* = *Longidorus* sp. *Meli* = *Meloidogyne incognita*. *Melm* = *Meloidogyne marioni*. *Meln* = *Meloidogyne naasi*. *Mesb* = *Mesodorylaimus bastiani*. *Monf* = *Mononchoides fortidens*. *Monl* = *Mononchoides longicaudatus*. *Mono* = *Mononchus aquaticus*. *Mym* = *Mylonchulus minor*. *Pan* = *Panagrellus redivivus*. *Parc* = *Paralongidorus citri*. *Pars* = *Paratrichodorus* sp. *Ple* = *Plectus* sp. *Ras* = *Radopholus similis*. *Rhs* = *Rhabditis* sp. *Rore* = *Rotylenchulus reniformis*. *Roro* = *Rotylenchus robustus*. *Scs* = *Scutellonema* sp. *Trs* = *Trichodorus* sp. *Tylm* = *Tylenchorhynchus mashhoodi*. *Tyls* = *Tylenchulus semipenetrans*. *Xia* = *Xiphinema americanum*. *Xib* = *Xiphinema basiri*. *Xii* = *Xiphinema insigne*. *Xis* = *Xiphinema* sp.

Table 11.1. *(Continued.)*

[2] Study type: B = Feeding behaviour. C = Cultured nematodes feeding on added prey. D = Direct observation of feeding process. E = Feeding experiment. G = Gut content observation. M = Molecular analysis (PCR) of gut content. T = Tentative assumption.
[3] References: 1 = Linford (1937). 2 = Linford & Oliveira (1937). 3 = Goodey (1951). 4 = Goodey (1963). 5 = Yeates (1971). 6 = Vinciguerra & Zullini (1978). 7 = Esser (1987). 8 = Small (1987). 9 = Ferris & Ferris (1989). 10 = Hunt (1978). 11 = Khan *et al.* (1995b). 12 = Khan & Kim (2007). 13 = Small & Grootaert (1983). 14 = Cabos *et al.* (2013). 15 = Khan & Jairajpuri (1997). 16 = Mankau (1980). 17 = Khan *et al.* (1991). 18 = Bilgrami (1993). 19 = Pervez (2008). 20 = Thorne (1928). 21 = Thorne & Swanger (1936). 22 = Thorne (1929). 23 = Tjepkema *et al.* (1971). 24 = Steiner (1924). 25 = Thorne (1930). 26 = Doncaster (1962). 27 = Wood (1973). 28 = Sohlenius (1979). 29 = Khan *et al.* (1995a). 30 = Bilgrami (2008). 31 = Marinari *et al.* (1982). 32 = Thorne (1939). 33 = Yeates (1967b). 34 = Andrássy (1969). 35 = Shafqat *et al.* (1987). 36 = Bilgrami (1992). 37 = Cobb (1929). 38 = de Man (1884). 39 = Nielsen (1949). 40 = Ferris (1968). 41 = Pillai & Taylor (1968). 42 = Russell (1986). 43 = Wyss & Grootaert (1977). 44 = Grootaert & Small (1982). 45 = Esser (1963). 46 = Small (1979). 47 = Pervez & Bilgrami (2000b). 48 = Bilgrami & Gaugler (2005). 49 = Bilgrami (1995). 50 = Bilgrami & Pervez (2000). 51 = Pervez & Bilgrami (2000a). 52 = Pervez *et al.* (2000). 53 = Bilgrami & Liang (2004). 54 = Spaull (1973). 55 = Bilgrami *et al.* (1985). 56 = Yeates (1967a). 57 = Thorne (1961). 58 = Hollis (1957). 59 = Boosalis & Mankau (1965). 60 = Yeates (1968). 61 = Micoletzky (1921). 62 = Micoletzky (1925). 63 = Sohlenius *et al.* (1977). 64 = Okada *et al.* (2005).
[4] Often recorded as *Aporcelaimellus obscurus*, its synonym.
[5] Also recorded as *Eudorylaimus irritans*.
[6] Recorded as *Durinema gracile*, its synonym.
[7] Occasionally recorded as *Discolaimus arenicolus*, its synonym.
[8] This very early observation by de Man (1884) of *Dorylaimus* sp. feeding on *Cephalobus* sp. should be taken with caution as at that time only a few dorylaimid species were known, and the precise identity of the predator remains uncertain.

It is an unquestionable fact that many dorylaimid forms are active predators (Esser, 1987; Small, 1987; Bilgrami, 1990, 2008; Khan & Kim, 2007). Table 11.1 presents an updated list of the species currently demonstrated to be able to predate on a variety of animal prey. It includes 46 species and 29 genera, a cohort dominated by representatives of the families Actinolaimidae, Aporcelaimidae and Nygolaimidae. The

Table 11.2. *Feeding sources for dorylaims, other than animal material.*

Taxon	Feeding source	Study type[1]	Original reference[2]	Later reference or compilation[2]
Actinolaimoides sp.	?Algae	G	4	
Aporcelaimellus obtusicaudatus	?Algae	C	39	6
Aporcelaimellus waenga	?Algae	G, T	56, 33	
Aporcelaimellus sp.	Algae, moss	C, D	27	
Aporcelaimus sp.	Algae, moss	C, D	27	
Aporcelinus americanus[3]	Algae, protozoans	C	9	
	Algae, protozoans-rhizopods		39	8
Aporcella taylori	?Algae	G, T	59, 33	
Crassolabium australe	?Algae	G, T	60, 56, 33	
Crassolabium ettersbergense	Algae, protozoa-cysts, fungi-conidia	C, D	58	4, 6, 8
Crocodorylaimus flavomaculatus	Moss	?D, G	61	39
	Algae	?D, G	62	39
Chrysodorus filiformis	Algae	?D, G	62	39
Discolaimium maitai	?Algae	G, T	56	
Dorylaimus sp.	Protozoans-ciliates		1	8
Enchodelus spp.	Plant-feeding	G	32	3, 4
Eudorylaimus carteri	Algae, moss	C	39	6
Eudorylaimus sp.	Fungi		59	8
	Algae, moss	C, D	27	
	Algae, fungi	C	7	
Labronema ferox	?Algae, ?moss	G	32	
Labronema hyalinum	Algae, moss	G	3	4
Labronema rikia	?Algae	G, T	56, 33, 60	
Labronema varicaudatum	Algae, moss	G	3	4
Labronema sp.	Fungi-spores	C, D	9	
	Fungi-spores	C, D	40	6, 8
Leptonchus sp.	Fungi-hyphae/spores	C, D	9	
	Algae, moss	C, D	27	
Mesodorylaimus cf. *lissus*	Protozoa-amoebae, fungi-hyphae, algae	C, D	42	

Table 11.2. *(Continued.)*

Taxon	Feeding source	Study type[1]	Original reference[2]	Later reference or compilation[2]
Mesodorylaimus cf. *pseudobastiani*	Algae, protozoans, ?fungi	C	9	
Mesodorylaimus sp.	Cocoid algae, fungi-spore	D	54	8
Meylis dicephalus	Plant-feeding	D, T	60	
Thornia sp.	Protozoa-amoeba, fungi, yeast	C	59	7, 8
Tylencholaimellus montanus	Fungi	C, D	27	
Tylencholaimus mirabilis	?Algae	G	39	
	Mycorrhizal fungi	T	63	
Tylencholaimus parvus	Fungi	C, E	64	
Tylencholaimus stecki	Mycorrhizal fungi	T	63	
Tylencholaimus sp.	Fungi-hyphae	C, D	27	
	Fungi-hyphae	C, D	9	

[1] Study type: C = Cultured nematodes feeding on added materials. D = Direct observation of feeding process. E = Feeding experiment. G = Gut content observation. T = Tentative assumption.

[2] See references in Table 11.1.

[3] Recorded as *Eudorylaimus irritans*.

predacious abilities of these taxa were originally noted by means of direct observations (de Man, 1884; Thorne, 1928; Cobb, 1929; Linford & Oliveira, 1937), although this was based mainly on the examination of the gut contents of fixed specimens (compiled by Thorne, 1939), and later corroborated by experimental essays. Dorylaims may feed on several kinds of invertebrate prey, mainly micro-oligochaetes (tubificid and enchytraeid forms), but also rotifers and mites. Nevertheless, cultures using other nematodes as prey have been especially successful, and the list of possible prey species is long, currently comprising a total of 52 (compiled in the corresponding column of Table 11.1).

Some predacious species, in particular the nygolaims (Nygolaimidae), are 'ingesters', able to capture and ingest entire prey, but the majority, other than nygolaims, are 'piercers' that suck the body fluids of their prey through their odontostyle (Yeates, 1971; Vinciguerra & Zullini,

1978; Yeates *et al.*, 1993). The predatory ability of dorylaims for use in biological control of plant-parasitic nematodes has been extensively evaluated and discussed, although without totally satisfactory results and conclusions (Small, 1979; Small & Grootaert, 1983; Esser, 1987; Bilgrami & Gaugler, 2005; Khan & Kim, 2007; Devi & George, 2018). Yet, the predation ability of dorylaims probably does not compare to that of the mononchs, apparently due to their lower mobility and/or speed of reaction and the need to rupture the prey cuticle with the odontostyle, rather than engulfing the entire prey (Small & Grootaert, 1983; but see Bilgrami & Gaugler, 2005).

Herbivory or plant-feeding, understood as vascular plant ecto-parasitism, is the feeding habit of longidorid forms, the representatives of the very important family Longidoridae, some of whose species are able to transmit plant viruses. Particular genera of other families, for example, *Californidorus, Enchodelus* and *Longidorella* in Nordiidae, which are armed with a long and slender odontostyle, are also suspected of feeding on vascular plants (*cf.* Thorne, 1939; Vinciguerra & Zullini, 1978; Yeates *et al.*, 1993). All these dorylaimid forms are piercers, puncturing the plant cell wall with the odontostyle and sucking out the plant fluids by the pumping action of their pharynx (see below) and probably aided by turgor pressure in the plant cell.

Of special mention is the case of mosses as a feeding resource. They have been used as a substrate to culture members of the genera *Aporcelaimellus*, *Aporcelaimus*, *Crocodorylaimus*, *Eudorylaimus*, *Labronema*, and *Leptonchus*, often in combination with algae (Table 11.1). Feeding on mosses might be regarded as a special case of plant-feeding.

There is less evidence of dorylaims as consumers of fungi (fungivory), although it does occur. Species of the genera *Crassolabium*, *Eudorylaimus*, *Labronema*, *Leptonchus*, *Mesodorylaimus*, *Thornia*, *Tylencholaimellus*, and *Tylencholaimus* have been successfully cultured using fungal materials (hyphae, spores, conidia) or have been observed feeding on them (see compilation in Table 11.2). In this regard, the experimental demonstration by Okada *et al.* (2006), working on *Tylencholaimus parvus*, is relevant. It does not seem irrational to conclude that fungivorous dorylaims are basically piercers, with a basic strategy similar to ecto-parasitic forms.

Microherbivory, which is the feeding on unicellular eukaryotic organisms, mainly algae (algivory), but also yeasts and protozoa, also occurs in dorylaims. Table 11.2 presents a long list of taxa cultured on substrates containing such organisms. In addition, these feeding sources seem to be naturally used in the wild by some particular genera. Thus, the intestine of *Rhyssocolpus* and *Heterodorus* species (Nordiidae) very often becomes distinctly greenish due to the abundant green material filling the lumen, a material that presumably comes from algae (*cf.* Yeates *et al.*, 1993). Also, some experimental studies have demonstrated that representatives of other genera, *e.g.*, *Aporcelaimellus*, may be cultured for long periods when feeding only on a few species of algae (Nielsen, 1949; Wood, 1974).

Some dorylaims may be able to use several feeding resources and hence practise omnivory. A comparison of Table 11.1 and 11.2 readily reveals that many genera, and even species, are included in both tables. This means that there is observational and/or experimental evidence that these taxa are potential omnivorous forms, combining animal and other materials in their diet. The taxa listed in Table 11.2 were often cultured using mixed substrates. Exactly to what extent omnivory occurs amongst dorylaims is, however, an interesting issue for further studies.

It is entirely possible that dorylaims exploit other types of food resource. Russell (1986) mentioned that survival and reproduction of a *Mesodorylaimus* species was successful when feeding on bacteria. Although no experimental demonstration has been provided so far, the possibility that some dorylaimid species may ingest organic matter in decomposition should not be dismissed. For example, it is not unusual to observe dark (brownish or blackish) material filling the intestine of specimens belonging to the genus *Aporcelaimellus*, which might be ingested deliberately.

Transcuticular uptake of nutrients, *via* cuticle body pores and canals, might be more common in dorylaims than in other nematodes (Hollis & Jordan, 1962; McNamara, 1980) and has been experimentally demonstrated in longidorid species (Mayo & Thomas, 1971).

Finally, there are several, almost anecdotal reports of dorylaims (?) dwelling inside the body cavities of animals. Thus, Dujardin (1845) collected *Dorylaimus stagnalis*, the first described dorylaimid species, from the stomach of the fishes *Cyprinus carpio* and *Gasterosteus laevis*, and Navarro *et al.* (1995) originally described *Dorylaimus parasiticus* on the basis of specimens collected from the gut of amphibians. It is possible, and even probable, that these nematodes were ingested together with other materials or as a food source.

Nevertheless, more intriguing is the reported occurrence of three species (*Aporcelaimellus* sp., *Discolaimus* sp. and *Ecumenicus monohystera*) within the haemocoel of insects (Namjou *et al.*, 2013).

Feeding behaviour

The feeding behaviour of dorylaims has been studied experimentally in several species. The predatory behaviour pattern is probably the most complex process and comprises five phases (Khan *et al.*, 1991; Bilgrami, 2008): predator-prey encounter, attack response, attack, salivation/extra-corporal digestion, and ingestion. Grootaert & Wyss (1978; see also Linford & Oliveira, 1937) explained many details of the predator behaviour of a *Labronema* species feeding on a rhabditid prey, which are summarised in the following[2]. After labial contact with the prey, the odontostyle penetrates the prey cuticle. Immediately after penetration the contractions of the pharyngeal musculature provoke a sucking action, apparently without any secretions from the pharyngeal glands being injected into the prey. The movement of the odontostyle is directed by the system of protractor and retractor muscles and involves the compression of a hydrostatic cylinder, *i.e.*, the space between the guiding sheath and the pharyngeal epidermis (Fig. 11.1A). The ingestion process takes place with the odontostyle totally protracted. A typically intermittent pumping causes a strong suction force to enable ingestion of the food. The lumen of the pharynx also plays an active role in the process as its opening and closing positions (Fig. 11.1B) facilitate the food flow. A rather similar behaviour was also observed in species feeding on algae (Wood, 1974) and in other predators (Small & Grootaert, 1983). The predation ability of dorylaims may be greater than is often assumed. Thus, Esser (1987) wrote (p. 3): "One hour after adding 500 specimens of *Paratylenchus curvitatus* to a culture containing mixed dorylaim genera many were succumbing to attack. In 3 days there were only a few survivors. Two days after adding 1000 *Pratylenchus penetrans* to a culture containing several genera of dorylaims all had been devoured, 176 more were added in the afternoon, none of which were alive the next morning". Chemosensory responses have been demonstrated by several

[2] Grootaert & Wyss (1978) also found differences in the functional anatomy of the anterior feeding apparatus between different taxa, for example free-living and plant-parasitic forms.

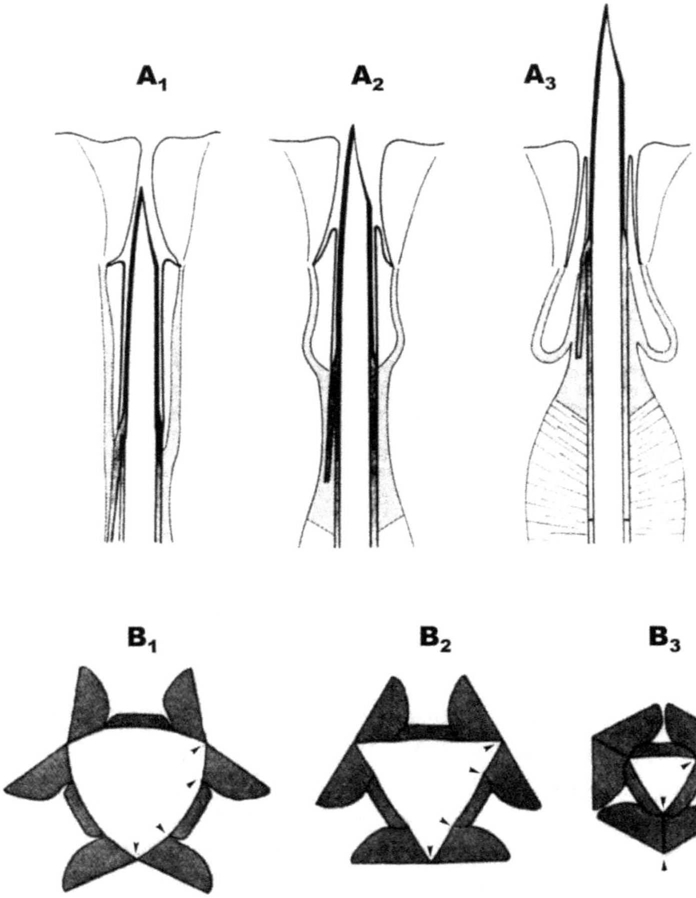

Fig. 11.1. *Some aspects of feeding behaviour in dorylaims. A: Schematic representation of compression of 'hydrostatic cylinder' and folding of guiding sheath during odontostyle protraction, with retracted (A_1), partially protruded (A_2) and totally protruded odontostyle (A_3) (after Grootaert & Wyss (1978),* Nematologica *24, p. 246, Fig. 8); B: Schematic representation of opening and closing of lumen within slender anterior portion of pharynx, with fully dilated lumen (B_1), lumen starting to narrow (B_2), and greatest possible narrowing (B_3) (arrowheads indicate bending points of cuticle). (After Grootaert & Wyss (1978),* Nematologica *24, p. 248. Fig. 11.)*

predator nematodes towards kairomones emitted by the prey (Bilgrami & Pervez, 2000; Pervez & Bilgrami, 2000a, b; Pervez, 2008), with different degrees of attraction depending on prey type (epidermal feeders better than virus vectors) and several factors (prey density, starvation, temperature, *etc.*). Bilgrami (1992) explored the predation and strike rate of *Dorylaimus stagnalis* and *Aquatides aquaticus* and found that all saprophagous forms were highly susceptible to predation, but also that some prey resisted predation by chemical means (toxic secretions), that dorylaimid and trichodorid prey are provided with physical (cuticular) features that provided partial resistance against predation, and that some species (hoplolaims and *Hemicycliophora* spp.) were totally resistant to predation.

The feeding behaviour of plant-parasitic forms has been studied in some detail, and apparently consists of two types of ingestion: an intermittent pumping of the basal bulb, and a continuous pumping, the latter occurring only at deep penetration and lasting for a considerable time (Hunt & Towle, 1979). Wyss *et al.* (1988; see also Wyss, 1975) and Trudgill *et al.* (1991) provided relevant and precise information about *Xiphinema index* and *X. diversicaudatum* behaviour, respectively. They fed on a column of progressively deeper plant cells and, following penetration of a root cell wall, secretions from the dorsal gland were injected whilst fluid, probably from the subventral glands, passed posteriad through the pharyngo-intestinal valve. This was followed by several periods of pumping separated by short pauses, at the beginning of which further secretions from the dorsal gland occurred.

Food suitability and selection

These aspects of the feeding biology of dorylaims have been little explored, but there are some interesting observations on the matter. Cell size may condition the suitability of some unicellular algae for some dorylaim species feeding on this type of resource. Thus, *Aporcelaimellus* sp. failed to penetrate small cells less than about 10 μm in diameter (Wood, 1974). The nature of the cuticle is probably a determinant in the suitability of many nematode species as potential prey for predatory dorylaims, those forms having a thick cuticle with a gelatinous basal layer being largely resistant to attack by predators (Small & Grootaert, 1983; Bilgrami, 1993, 1995). The release of repulsive/non-attractive secretions by the potential prey may also result in the prevention of attack by predators (Bilgrami, 1993, 1995).

Polyphagy seems to be a general pattern of the predator/prey relationship since there is no documented case of a predacious dorylaim feeding exclusively on one prey species. Interesting mathematical models have been developed to characterise the many aspects of predator/prey relationships (Bilgrami *et al.*, 1985; Khan *et al.*, 1991; Bilgrami, 1993). Bilgrami & Gaugler (1985) pointed out that the prey search and killing abilities of two dorylaimid species were governed by temperature, prey density, starvation and prey incubation, and depended upon feeding duration, number of predators feeding and aggregation at the feeding sites.

Feeding functional morphology

The basic structure of stoma and pharynx may play an important role and determine the feeding behaviour of dorylaims. Thus, Geraert (1978) stated that a fine stomatal lumen (allowing only liquid food) is apparently associated with a slender, non-muscular, anterior part of the pharynx followed by a cylindrical, muscular basal bulb, whereas a wider stomatal lumen (allowing particulate food) is apparently associated with a muscular, anterior region of the pharynx. Many details of the action of the pharyngeal feeding pump was provided by Seymour (1983) and Robertson *et al.* (1987) in *Longidorus* species, with a discussion of the nature of closing forces, function of the lining plates, sequential control of pumping, and transition zones between the open and closed pump lining. In addition, Tahseen *et al.* (2004) studied the physiology (contraction and relaxation) of the pharyngeal musculature in *Aporcelaimus superbus* by mean of electropharyngeograms and found only occasional electrical activity when this species was exposed to serotonin and no stylet activity in any treatment.

Excretion and osmo-regulation in dorylaims

Available information about the excretory organs in dorylaims is extremely poor. Only a rudimentary excretory pore is sometimes traceable, and it is very often assumed that this kind of organ is totally absent. Aboul-Eid (1969), however, described with some detail the excretory system in *Longidorus macrosoma* (Fig. 11.2) although the

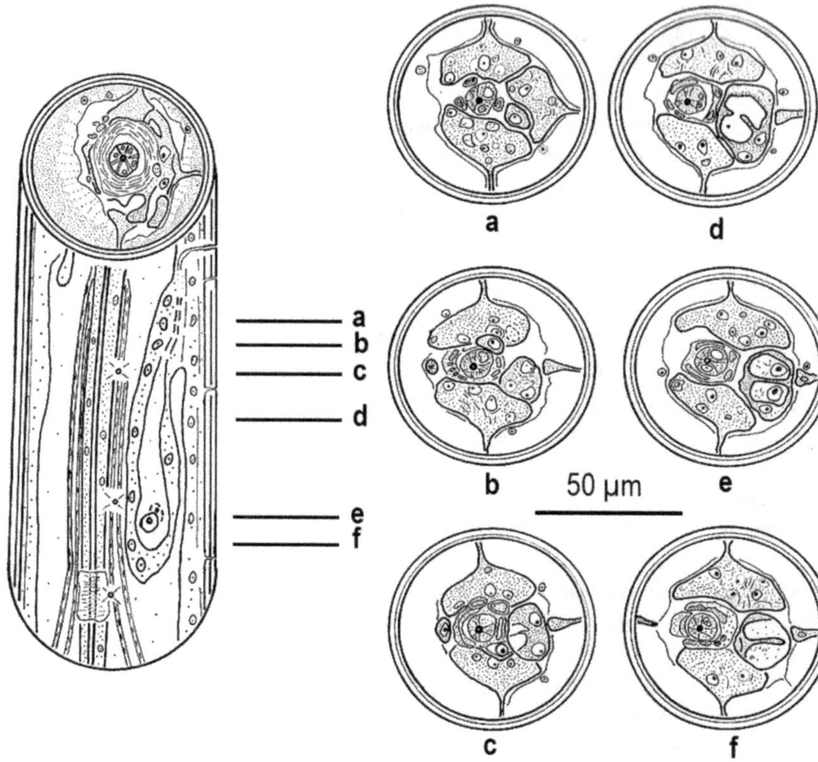

Fig. 11.2. *Schematic drawing of the excretory system of* Longidorus macrosoma. *a-f: Transverse sections in the body region occupied by the excretory system. (After Aboul-Eid (1969),* Nematologica *15, p. 438, Fig. 1.)*

actual excretory function of this structure was not proved. The excretory system would tentatively consist of three components: two nucleate gland cells embedded in ganglia tissue, an ampulla-like structure located immediately posterior to the level of the excretory pore, and an excretory pore opening ventrally at nerve ring level.

References

ABOUL-EID, H.Z. (1969). Histological anatomy of the excretory and reproductive systems of *Longidorus macrosoma*. *Nematologica* 15, 437-450. DOI: 10.1163/187529269X00777

ANDRÁSSY, I. (1969). Taxonomische Übersicht der Familien Prodorylaimidae n. fam. und Dorylaimidae de Man, 1876. *Opuscula Zoologica Budapestinensis* 9, 187-233.

BILGRAMI, A.L. (1990). Predaceous dorylaims. In: Jairajpuri, M.S., Mashkoor, M. & Ahmad, I. (Eds). *Nematode bio-control: aspects and prospects*. New Delhi, India, CBS Publishers and Distributors Pvt Ltd, pp. 143-148.

BILGRAMI, A.L. (1992). Resistance and susceptibility of prey nematodes to predation and strike rate of the predators *Mononchus aquaticus*, *Dorylaimus stagnalis* and *Aquatides thornei*. *Fundamental and Applied Nematology* 15, 265-270.

BILGRAMI, A.L. (1993). Analysis of the predation by *Aporcelaimellus nivalis* on prey nematodes from different prey trophic categories. *Nematologica* 39, 356-365. DOI: 10.1163/187529293X00303

BILGRAMI, A.L. (1995). Numerical analysis of the predatory relations between *Mesodorylaimus bastiani* (Nematoda: Dorylaimida) and different prey trophic categories. *Nematologia Mediterranea* 2, 81-88.

BILGRAMI, A.L. (2008). Biological control potentials of predatory nematodes. In: Ciancio, A. & Mukerji, K.J. (Eds). *Integrated management and biocontrol of vegetable and grain crops nematodes*. Berlin, Germany, Springer, pp. 3-28.

BILGRAMI, A.L. & GAUGLER, R. (2005). Feeding behaviour of the predatory nematodes *Laimydorus baldus* and *Discolaimus major* (Nematoda: Dorylaimida). *Nematology* 7, 11-20. DOI: 10.1163/1568541054192207

BILGRAMI, A.L. & LIANG, W. (2004). Effect of temperature, predator starvation, prey density and prey incubation on feeding and aggregation activities of predatory nematodes, *Mesodorylaimus bastiani* and *Aquatides thornei*. *Nematologia Mediterranea* 32, 87-94.

BILGRAMI, A.L. & PERVEZ, R. (2000). Prey searching and attraction behaviour of *Mesodorylaimus bastiani* and *Aquatides thornei* (Nematoda: Dorylaimida). *International Journal of Nematology* 10, 199-206.

BILGRAMI, A.L., AHMAD, I. & JAIRAJPURI, M.S. (1985). Predatory behaviour of *Aquatides thornei* (Nygolaimina: Nematoda). *Nematologica* 30(1984), 457-462. DOI: 10.1163/187529284X00356

BOOSALIS, M.G. & MANKAU, R. (1965). Parasitism and predation of soil microorganisms. In: Baker, K.F. & Sneider, W.C. (Eds). *Ecology of soil-borne plant pathogens*. Berkeley, CA, USA, University of California Press, pp. 374-391.

CABOS, R.Y.M., WANG, K.-H., SIPES, B.S., HELLER, W.P. & MATSUMOTO, T.K. (2013). Detection of plant-parasitic nematode DNA in the gut of predatory and omnivorous nematodes. *Nematropica* 43, 44-48.

COBB, N.A. (1929). Nemas of the genus *Dorylaimus* attacking the eggs of mites. *Journal of Parasitology* 15, 284.

de Man, J.G. (1884). *Die frei in der reinen Erdee und im süssen Wasser lebenden Nematoden der niederländischen Fauna. Eine systematische-faunistiche Monographie*. Leiden, The Netherlands, E.J. Brill.

Devi, G. & George, J. (2018). Predatory nematodes as bio-control agent against plant-parasitic nematode – a review. *Agricultural Reviews* 39, 55-61. DOI: 10.18805/ag.r-1715

Doncaster, C.C. (1962). Predators of soil nematodes (film). *Parasitology* 52, 19P.

Dujardin, F. (1845). *Histoire naturelle des helminthes ou vers intestinaux*. Paris, France, Librairie Encyclopédique de Roret. DOI: 10.5962/bhl.title.10123

Esser, R.P. (1963). Nematode interactions in plates of non-sterile water agar. *Proceedings of the Soil and Crop Science Society of Florida* 23, 121-138.

Esser, R.P. (1987). Biological control of nematodes by nematodes. I. Dorylaims (Nematoda: Dorylaimina). *Nematology Circulars, Florida Department of Agriculture and Consumer Service, Division of Plant Industry* 144, 1-4.

Ferris, V.R. (1968). Biometric analyses in the genus *Labronema* (Nematoda: Dorylaimida) with a description of *L. thornei* n. sp. *Nematologica* 14, 276-284. DOI: 10.1163/187529268X00480

Ferris, V.R. & Ferris, J.M. (1989). Why ecologists need systematics: Importance of systematics to ecological research. *Journal of Nematology* 21, 308-314.

Geraert, E. (1978). On growth and form in nematodes: II. Oesophagus and body width in Dorylaimida. *Nematologica* 24, 347-360. DOI: 10.1163/187529278X00470

Goodey, J.B. (1963). *Soil and freshwater nematodes*, 2nd edition. London, UK, Methuen & Co., Ltd.

Goodey, T. (1951). *Soil and freshwater nematodes – a monograph*. London, UK, Methuen & Co., Ltd.

Grootaert, P. & Small, R.W. (1982). Aspects of the biology of *Labronema vulvapapillatum* (Meyl) (Nematoda: Dorylaimida) in laboratory culture. *Biologisch Jaarboek Dodonaea* 50, 135-148.

Grootaert, P. & Wyss, U. (1978). Ultrastructure and function of the anterior feeding apparatus in *Labronema* sp. *Nematologica* 24, 243-250. DOI: 10.1163/187529278X00209

Hollis, J.P. (1957). Cultural studies with *Dorylaimus ettersbergensis*. *Phytopathology* 47, 468-473.

Hollis, J.P. & Jordan, J.H. (1962). Differential penetrability of nematodes. *Phytopathology* 52, 14.

Hunt, D.J. (1978). Bionomics of *Paractinolaimus vigor* Thorne, 1967 (Dorylaimida: Paractinolaimidae) with a description of *P. dominicus* n. sp. and

Nygolaimium haguei n. sp. (Dorylaimida: Aporcelaimidae). *Nematologica* 23(1977), 452-462. DOI: 10.1163/187529277X00381

HUNT, D.J. & TOWLE, A. (1979). Feeding studies on *Xiphinema vulgare* Tarjan, 1964 (Nematoda: Longidoridae). *Revue de Nématologie* 2, 37-40.

KHAN, Z. & JAIRAJPURI, M.S. (1997). Predatory behaviour of *Paractinolaimus elongatus* on plant parasitic nematodes. *International Journal of Nematology* 7, 178-181.

KHAN, Z. & KIM, Y.H. (2007). A review on the role of predatory soil nematodes in the biological control of plant parasitic nematodes. *Applied Soil Ecology* 35, 370-379. DOI: 10.1016/j.apsoil.2006.07.007

KHAN, Z., BILGRAMI, A.L. & JAIRAJPURI, M.S. (1991). Observations on the predation ability of *Aporcelaimellus nivalis* (Altherr, 1952) Heyns, 1986 (Nematoda: Dorylaimida). *Nematologica* 37, 333-342. DOI: 10.1163/187529291X00330

KHAN, Z., BILGRAMI, A.L. & JAIRAJPURI, M.S. (1995a). A comparative study on the predation by *Allodorylaimus americanus* and *Discolaimus silvicolus* (Nematoda: Dorylaimida) on different species of plant parasitic nematodes. *Fundamental and Applied Nematology* 18, 99-108.

KHAN, Z., BILGRAMI, A.L. & JAIRAJPURI, M.S. (1995b). Observations on the predation abilities of *Neoactinolaimus agilis* (Dorylaimida: Actinolaimoidea). *Indian Journal of Nematology* 25, 129-135.

LINFORD, M.B. (1937). The feeding of some hollow stylet nematodes. *Proceedings of the Helminthological Society of Washington* 4, 41-46.

LINFORD, M.B. & OLIVEIRA, J.M. (1937). The feeding of hollow spear nematodes on other nematodes. *Science* 85, 295-297. DOI: 10.1126/science.85.2203.295

MANKAU, R. (1980). Biological control of nematode pests by natural enemies. *Annual Review of Phytopathology* 18, 415-440. DOI: 10.1146/annurev.py.18.090180.002215

MARINARI, A., VINCIGUERRA, M.T., VOVLAS, N. & ZULLINI, A. (1982). Nematodi delle dune costiere d'Italia. In: *Struttura delle zoocenosi terrestri 3. Ambienti Mediterranei I. Le Coste Sabbiose*. Rome, Italy, Consiglio Nazionale delle Ricerche, pp. 27-50.

MAYO, M.A. & THOMAS, P.R. (1971). Labelling of *Longidorus elongatus* and *Xiphinema diversicaudatum* with glucose [^{14}C]. *Nematologica* 17, 23-28. DOI: 10.1163/187529271X00378

MCNAMARA, D.G. (1980). The survival of *Xiphinema diversicaudatum* in plant-free soil. *Nematologica* 26, 170-181. DOI: 10.1163/187529280X00062

MICOLETZKY, D.H. (1922). Die freilebenden Erd-nematoden. *Archiv für Naturgeschichte* 87(1921), 1-650.

MICOLETZKY, D.H. (1925). Die freilebenden Süsswasser- und Moornematoden Dänemarks. *Mémoires de l'Académie Royale des Sciences et des Lettres de Danemark, Copenhague* 10, 57-310.

NAMJOU, S., PEDRAM, M., POINAR JR, G.O., POURJAM, E. & ATIGHI, M.R. (2013). Novel host associations for dorylaim nematodes (Nematoda: Dorylaimida). *International Journal of Nematology* 23, 111-114.

NAVARRO, P., GUERRERO, F., PÉREZ-MELLADO, V. & LLUCH, J. (1995). Description of a new species: *Dorylaimus parasiticus*, a parasite of amphibians in the Iberian Peninsula (Nematoda: Dorylaimida). *Journal of Zoology* 237, 160-177. DOI: 10.1111/j.1469-7998.1995.tb02756.x

NIELSEN, C.O. (1949). Studies on the soil microfauna. II. The soil inhabiting nematodes. *Natura Jutlandica* 2, 1-131.

OKADA, H., HARADA, H., TSUKIBOSHI, T. & ARAKI, M. (2006). Characteristics of *Tylencholaimus parvus* (Nematoda: Dorylaimida) as a fungivorus nematode. *Nematology* 7(2005), 843-849. DOI: 10.1163/156854105776186424

PERVEZ, R. (2008). Chemosensory responses of *Aporcelaimellus nivalis* (Nematoda: Dorylaimida) towards kairomones emitted by prey nematodes belonging to different trophic groups. *Archives of Phytopathology and Plant Protection* 41, 318-327. DOI: 10.1080/03235400600759404

PERVEZ, R. & BILGRAMI, A.L. (2000a). Chemosensory responses of predatory nematodes towards kairomones emitted by prey nematodes belonging to different trophic groups. *Nematologia Mediterranea* 28, 267-277.

PERVEZ, R. & BILGRAMI, A.L. (2000b). Some factors influencing chemoattraction behavior of dorylaim predators, *Laimydorus baldus* and *Discolaimus major* towards prey kairomones. *International Journal of Nematology* 10, 41-48.

PERVEZ, R., BILGRAMI, A.L., YOSHIGA, T. & KONDO, E. (2000). Feeding, attraction and aggregation behaviour of *Mesodorylaimus bastiani* and *Aquatides thornei* at the feeding site using *Hirschmanniella oryzae* as prey. *International Journal of Nematology* 10, 207-214.

PILLAI, J.K. & TAYLOR, D.P. (1968). Biology of *Paroigolaimella bernensis* and *Fictor anchicoprophaga* (Diplogasterinae) in laboratory culture. *Nematologica* 14, 159-170. DOI: 10.1163/187529268X00381

ROBERTSON, W.M., TOPHAM, P.B. & SMITH, P. (1987). Observations on the action of the oesophageal pump in *Longidorus* (Nematoda). *Nematologica* 33, 43-54. DOI: 10.1163/187529287X00191

RUSSELL, C.C. (1986). The feeding habits of a species of *Mesodorylaimus*. *Journal of Nematology* 18, 641.

SEYMOUR, M.K. (1983). Action of the feeding pump in *Longidorus caespiticola* Hooper. *Nematologica* 29, 65-77. DOI: 10.1163/187529283X00195

SHAFQAT, S., BILGRAMI, A.L. & JAIRAJPURI, M.S. (1987). Evaluation of the predation ability of *Dorylaimus stagnalis* Dujardin, 1845 (Nematoda: Dorylaimida). *Revue de Nématologie* 10, 455-461.

SMALL, R.W. (1979). The effects of predatory nematodes on populations of plant parasitic nematodes in pots. *Nematologica* 25, 94-103. DOI: 10.1163/187529279X00424

SMALL, R.W. (1987). A review of the prey of predatory soil nematodes. *Pedobiologia* 30, 179-206.

SMALL, R.W. & GROOTAERT, P. (1983). Observations on the predation abilities of some soil dwelling predatory nematodes. *Nematologica* 29, 109-118. DOI: 10.1163/187529283X00230

SOHLENIUS, B. (1979). A carbon budget for nematodes, rotifers and tardigrades in a Swedish coniferous forest soil. *Holartic Ecology* 2, 30-40.

SPAULL, V.W. (1973). Distribution of nematode feeding groups at Signy Island, South Orkney Islands with an estimate of their biomass and oxygen consumption. *British Antarctic Survey Bulletin* 37, 21-32.

STEINER, G. (1924). On some plant parasitic nemas and related forms. *Journal of Agriculture Research* 28, 1058-1079.

TAHSEEN, Q., SHERIDAN, J.P. & PERRY, R.N. (2004). Electropharyngeograms of pharyngeal pumping activity in six species of free-living nematodes. *Nematology* 6, 49-54. DOI: 10.1163/156854104323072919

THORNE, G. (1928). Nematodes inhabiting the cysts of the sugarbeet nematodes (*Heterodera schachtii* (Schmidt)). *Journal of Agriculture Research* 37, 571-575.

THORNE, G. (1929). Nematodes from the summit of Long's peak, Colorado. *Transactions of the American Microscopy Society* 48, 181-195.

THORNE, G. (1930). Predaceous nemas of the genus *Nygolaimus* and a new genus *Sectonema*. *Journal of Agricultural Research, USDA* 41, 445-466.

THORNE, G. (1939). A monograph of the nematodes of the superfamily Dorylaimoidea. *Capita Zoologica* 8, 1-261.

THORNE, G. (1961). *Principles of nematology*. New York, NY, USA, McGraw-Hill.

THORNE, G. & SWANGER, H.H. (1936). A monograph of the nematode genera *Dorylaimus* Dujardin, *Aporcelaimus* n. gen., *Dorylaimoides* n. gen., and *Pungentus* n. gen. *Capita Zoologica* 6, 1-223.

TJEPKEMA, J.P., FERRIS, V.R. & FERRIS, J.M. (1971). Review of the genus *Aporcelaimellus* Heyns, 1965 and six species groups of the genus *Eudorylaimus* Andrássy, 1959 (Nematoda: Dorylaimida). *Purdue University, Agriculture Experimental Station Research Bulletin* 882, 1-52.

TRUDGILL, D.L., ROBERTSON, W.M. & WYSS, U. (1991). Analysis of the feeding of *Xiphinema diversicaudatum*. *Revue de Nématologie* 14, 107-112.

VINCIGUERRA, M.T. & ZULLINI, A. (1978). Alimentazione e valenza ecologica dei nematodi liberi. *Nematologia Mediterranea* 6, 29-48.

WOOD, F.H. (1973). Nematode feeding relationships. Feeding relationships of soil-dwelling nematodes. *Soil Biology and Biochemistry* 5, 593-601. DOI: 10.1016/0038-0717(73)90049-7

WOOD, F.H. (1974). Biology of *Aporcelaimellus* sp. (Nematoda: Aporcelaimidae). *Nematologica* 19(1973), 528-537. DOI: 10.1163/187529273X00547

WYSS, U. (1975). Feeding of *Trichodorus, Longidorus* and *Xiphinema*. In: Lamberti, F., Taylor, C.E. & Seinhorst, J.W. (Eds). *Nematode vectors of plant viruses*. London and New York, Plenum Press and NATO Scientific Affairs Division, pp. 203-221.

WYSS, U. & GROOTAERT, P. (1977). Feeding mechanism in *Labronema* sp. *Mededelingen Faculteit Lanbouwwoheschool Rijksuniversiteit Gent* 42, 1521-1527.

WYSS, U., ROBERTSON, W.M. & TRUDGILL, D.L. (1988). Oesophageal bulb function of *Xiphinema index* and associated root cell responses, assessed by video-enhanced contrast light microscopy. *Revue de Nématologie* 11, 253-261.

YEATES, G.W. (1967a). Studies on nematodes from dune sands. 6. Dorylaimoidea. *New Zealand Journal of Science* 10, 752-784.

YEATES, G.W. (1967b). Studies on nematodes from dune sands. 9. Quantitative comparison of the nematode faunas of six localities. *New Zealand Journal of Science* 10, 927-948.

YEATES, G.W. (1968). An analysis of annual variation of the nematode fauna in dune sand at Himatangi Beach, New Zealand. *Pedobiologia* 8, 173-207.

YEATES, G.W. (1971). Feeding types and feeding groups in plant and soil nematodes. *Pedobiologia* 11, 173-179.

YEATES, G.W., BONGERS, T., DE GOEDE, R.G.M., FRECKMAN, D.W. & GEORGIEVA, S.S. (1993). Feeding habits in soil nematode families and genera – an outline for soil ecologists. *Journal of Nematology* 25, 315-331.

Nematology Monographs & Perspectives, 2021, Vol. 13, 171-192

Chapter 12

Reproduction and development

Bisexual *vs* monosexual species

Sexual reproduction is the only mode of reproduction known to occur in dorylaims with amphimixis the most common process, although parthenogenesis is also frequent. Conversely, there is nearly no evidence of hermaphroditism. The existence of two separate sexes (bisexual condition) is the usual scheme, which can occur in nearly two-thirds (62.7%) of the species. Nevertheless, bisexuality varies significantly depending on the considered taxon (Table 12.1). Thus, both female and male are present in all the members of a few genera, for example, *Labronemella* and *Lenonchium* (Table 12.1), whilst males remain totally unknown in the 13 *Drepanodorylaimus* species (Andrássy, 2009). Nevertheless, the existence of males does not ensure amphimixis. In their monograph on the very diverse genus *Xiphinema*, Coomans *et al.* (2001) pointed out that: *i*) in about 40% of the species males are present and functional, with reproduction occurring by amphimixis; *ii*) in 25% of the species males are rare[1] and reproduction probably takes place through parthenogenesis; and *iii*) in the remaining 35% of the species males are not known at all, reproduction occurring by obligate parthenogenesis.

Parthenogenesis is a derived condition that has certainly evolved independently many times at species level (Coomans *et al.*, 2001). It is also a successful reproduction strategy for K-strategists – and the dorylaims are certainly successful – that dwell in environmentally stable habitats. Meiotic parthenogenesis seems to be the general rule in the group (Dalmasso, 1975).

[1] Interestingly, Raschké & Boag (1981) reported a sex (female/male) ratio of 1: 0.014 in *Longidorus elongatus*.

Table 12.1. *Percentage of bisexual species in selected dorylaimid genera.*

Genus	n[1]	b[2]	m[3]	%[4]	Reference
Aporcelaimellus	28	14	14	50.0	Álvarez-Ortega & Peña-Santiago (2013a)
Aporcelaimoides	8	4	4	50.0	Álvarez-Ortega *et al.* (2015)
Aporcelinus	30	18	12	60.0	Peña-Santiago *et al.* (2019)
Aporcella	13	9	4	69.2	Álvarez-Ortega *et al.* (2013a)
Chitwoodius	12	5	7	41.7	Hoang *et al.* (2019)
Labronemella	9	9	0	100.0	Peña-Santiago *et al.* (2012)
Lenonchium	7	7	0	100.0	Asgari *et al.* (2020)
Makatinus	10	7	3	70.0	Peña-Santiago & Varela (2017)
Metaporcelaimus	30	13	17	43.3	Álvarez-Ortega *et al.* (2013b)
Metaxonchium	25	18	7	72.0	Peña-Santiago *et al.* (2014)
Nygolaimoides	7	6	1	85.7	Zahedi *et al.* (2016)
Oriverutus	22	12	10	54.5	Peña-Santiago *et al.* (2014)
Paravulvus	16	7	9	43.7	Gilarte *et al.* (2013)
Pungentus	16	8	8	50.0	Álvarez-Ortega & Peña-Santiago (2014)
Rhyssocolpus	8	7	1	87.5	Peña-Santiago *et al.* (2015)
Sectonema	26	18	8	69.2	Álvarez-Ortega & Peña-Santiago (2019)
Xiphinema	234	152	82	65.0	Coomans *et al.* (2001)
Total	501	314	187	62.7	

[1] n = total number of species in the corresponding genus.
[2] b = number of bisexual species, in which both females and males are known.
[3] m = monosexual species, with only females known.
[4] percentage of bisexual species.

Gametogenesis and egg production

Some details of these processes are well known in the case of a few longidorid forms that have been well studied as a result of the applied interest in their biology, although they remain nearly unexplored in the free-living taxa. Gametogenesis depends on environmental factors like climate or food. It takes place under favourable conditions and may decrease or cease under adverse situations (Dalmasso, 1975). The mating process requires the active contribution of both sexes and tactile stimuli are almost certainly involved (Trudgill, 1976). Reproduction may occur twice (February-April, September-October) a year, as observed in *Xiphinema basiri*, or only once (May-August) as happens in *X. insigne* (Jairajpuri & Bajaj, 1978). Flegg (1966, 1968a) also found that females of *X. vuittenezi* reproduce only once per year, and that they have more apparent and better developed ovaries between May and July.

Unfortunately, genetics remains a very poorly known aspect of dorylaimid biology. The chromosome number has been determined in longidorid taxa (Dalmasso, 1975; Kruger & Heyns, 1988), *n* being 7 in *Longidorus*, and 5 or 10 (2 × 5) in *Xiphinema* species. The cytogenetics was analysed in some detail by Dalmasso (1975), who found similar patterns of spermatogenesis and oogenesis in both *Longidorus* and *Xiphinema* species.

Regarding egg production, several studies have demonstrated that longidorid species only release eggs during a brief period in late spring and/or early summer (Flegg, 1968b; Thomas, 1969; Harris, 1979; Jaffee *et al.*, 1987). Field and culturing experiments with *X. diversicaudatum* suggest that the total reproductive capacity of this species shows significant variation, ranging from ten to 60 eggs per female each year under field conditions in southern England (Flegg, 1968b), and thrice this range under *in vitro* conditions (Brown & Coiro, 1983). Culturing conditions (host plant, experimental variables) can affect egg production. Thus, Brown & Coiro (1985) noted significant differences in *X. index*: an egg laid every 25 days at 10°C under fig host plants, but every 62 or 90 days under tomato host plants. The contribution of both genital tracts to egg production in didelphic species may be similar (equally functional), as observed in *X. basiri*, or not, with the anterior branch being much less efficient (ratio 1: 60) than the posterior one, as occurs in *X. insigne* (Jairajpuri & Bajaj, 1978).

Embryonic development and hatching

As is to be expected, the embryonic development and hatching process has been well documented in the plant-parasitic Longidoridae (for example, see Flegg, 1968c) and less so in economically less important free-living forms (for example, see Wood, 1974). Shafqat *et al.* (1991, 1994), however, studied these processes in *Dorylaimus stagnalis* and *Laevides symmetricus* in some detail (Fig. 12.1) and their findings for the former species are summarised as follows. Once mating and fertilisation have taken place, the single-celled egg divides perpendicularly to the long axis into two equal blastomeres 15-24 h after egg laying. Both cells subsequently divide following an oblique plane to the long axis. Subsequent divisions result in an eight-celled stage after 40 min, 12-

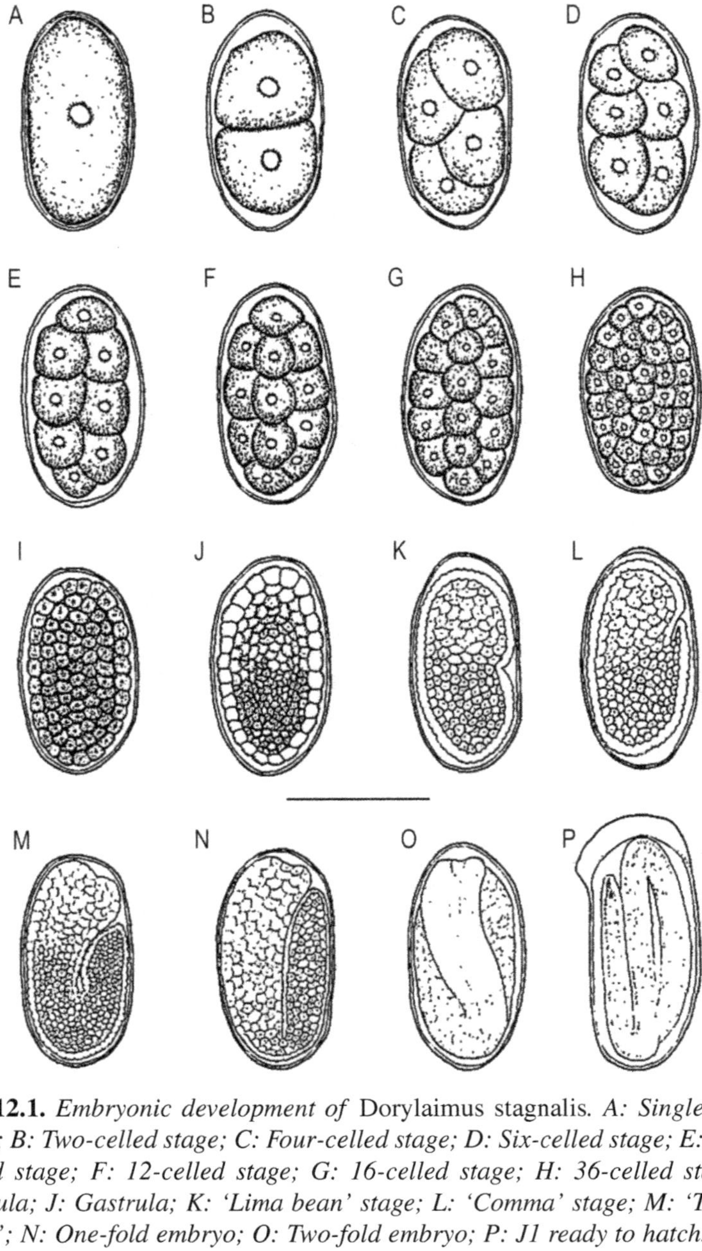

Fig. 12.1. *Embryonic development of* Dorylaimus stagnalis. *A: Single-celled stage; B: Two-celled stage; C: Four-celled stage; D: Six-celled stage; E: Eight-celled stage; F: 12-celled stage; G: 16-celled stage; H: 36-celled stage; I: Blastula; J: Gastrula; K: 'Lima bean' stage; L: 'Comma' stage; M: 'Tadpole stage'; N: One-fold embryo; O: Two-fold embryo; P: J1 ready to hatch. (After Shaftaq* et al. *(1991),* Revue de Nématologie *14, p. 63, Fig. 1.) (Scale bar = 50 μm.)*

celled after 45 min, 16-celled after about 1 h, and the 32-celled stage in the next 2 h[2]. Afterwards, the cells multiply rather rapidly.

The blastula appears 40-90 min later and the gastrula is formed 75-90 min after the formation of the blastula. The gastrula-forming cells differentiate into two types: an outer hyaline layer of cuboidal cells, and an inner portion forming a more granular mass. The development of the anterior region produces large cells from which originate the future stomodeum (feeding apparatus, pharynx), while the posterior region with smaller cells develops into the intestine and caudal region.

The gastrula becomes a 'lima bean' stage after 105 min by a slight invagination in the middle of the developing embryo and a 'comma' stage after another 75 min. The 'tadpole' stage appears 35 min later after the formation of a deep invagination. The mouth depression appears at the broader (anterior) end of the developing embryo in the late tadpole stage. Then the embryo significantly increases in length, by two-fold within 25 min and three-fold in another 160 min.

The different body regions are weakly perceptible after the three-fold stage is reached, and about 150 min later the tip of the functional odontostyle appears. The complete guiding ring, along with the intestine, appears about 4 h after the odontostyle tip is formed. The formation of a functional odontostyle takes 5 h, and is followed by the formation of the replacement odontostyle and the odontophore walls. The first-stage juvenile hatches from the eggs about 6 h after the formation of their internal organs. The egg shell becomes thin, flexible, increases in width, and forms a blister-like structure at the anterior region. After the formation of the blister, the continuous activity of the juvenile eventually breaks the egg membrane.

The total duration of the process, from single-cell stage to hatching took from 30 to 59 h (but see Table 12.2). Maximum hatching happened at 25-35°C, the maximum time required for hatching was 144 h at 15°C, the minimum 24 h at 40°C, and no hatching was observed at 5 and 10°C

[2] Shafqat *et al.* (1996), who studied the very early development in ten dorylaimid species, found that: *i*) the first division is transverse in all the species; *ii*) the second and third divisions were along an oblique longitudinal axis in nine species, but in one the second division was transverse and the third vertical; and *iii*) after the formation of four blastomeres, the main variation was from the fourth cleavage in the sequence of divisions between the five to eight cell stage.

Table 12.2. *Duration of embryonic development, life cycle and life span in selected dorylaimid species.*

Species	Embryonic development	Postembryonic development	Life span	Habitat	Reference
Aporcelaimellus sp.	10-12 days	95-130 days		Culture	Wood (1973)
Dorylaimus stagnalis	30-59 h	111-138 days		Culture	Shafqat *et al.* (1991)
Labronema ferox		3-4 months[1]		Culture	Ferris (1968)
Labronema thornei		5-6 months[1]		Culture	Ferris (1968)
Labronema vulvapapillatum	5-22 days	27-126 days[1]		Culture	Grootaert & Small (1982)
Laevides symmetricus	86-111 h			Culture	Shafqat *et al.* (1994)
Longidorus africanus		up to 4 months[1]		Culture (20-30°C)	Cohn & Mordechai (1969)
		8 weeks		Culture (tomato, 28°C)	Kolodge *et al.* (1987)
Longidorus athesinus	55-60 days[1]	26 weeks		Culture (wild cherry)	Coiro & Sasanelli (1995)
Longidorus elongatus	9-12 days	19 weeks[1]		Culture (strawberry, 20°C)	Yassin (1969)
		9 weeks[1]		Culture (strawberry, 30°C)	Wyss (1970)
Longidorus macrosoma	25-30 days			UK-England	Flegg (1968a)
Longidorus profundorum	25-30 days			UK-England	Flegg (1968a)
Paractinolaimus vigor	4 days	17-20 days		Culture	Hunt (1978)
Xiphinema basiri	*ca* 1 week			India	Jairajpuri & Bajaj (1978)
Xiphinema brevicollum		4-7 months[1]		Culture	Cohn & Mordechai (1969)

Table 12.2. *(Continued.)*

Species	Embryonic development	Postembryonic development	Life span	Habitat	Reference
Xiphinema diversicaudatum	19-24 days	3 years	5-6 years	UK-England	Flegg (1968a, b)
	12 weeks	60 weeks		Culture	Brown & Coiro (1983)
	12 weeks	56 weeks		Culture (*Lolium perenne*)	Coiro *et al.* (1999)
Xiphinema ifacolum	8 weeks	34 weeks		Culture	Coiro *et al.* (1995)
Xiphinema index		7-9 months[1]		Culture	Cohn & Mordechai (1969)
			12-14 months	Italy-Sardinia (vineyard)	Prota & Garau (1973)
		2-4 months		Culture	Prota *et al.* (1977)
		12 weeks		Culture	Coiro *et al.* (1991)
	12 days			Culture	Vovlas & Larizza (1994)
			64 weeks	Culture (fig)	Brown & Coiro (1985)
			40 weeks	Culture (tomato)	Brown & Coiro (1985)
		10 weeks		Culture (vineyard)	Coiro & Agostinelli (1991)
Xiphinema insigne	*ca* 1 week			India	Jairajpuri & Bajaj (1978)
Xiphinema vuittenezi	19-22 days	2-3 years	5-6 years	UK-England	Flegg (1968a, b)
		6 months[1]	24 months	Culture (vineyard)	Coiro *et al.* (2000)
Xiphinema vulgare		274 days[1]	20 months	Culture	Coiro *et al.* (2002)

[1] Life cycle (either from adult to adult or from egg to adult).

up to a period of 10 days. Boag (1985) studied the effect of temperature on hatching in *Longidorus elongatus* and found successful results at a range of 12.5-30°C.

Intra-uterine egg development and *endotokia matricida* (intra-uterine birth) is extremely exceptional in dorylaims, having being reported only in *Xiphinema* species (Jatala, 1975; Jairajpuri & Bajaj, 1978).

Postnatal development – general aspects

Dorylaims follow the typical nematode developmental pattern, basically consisting of four (exceptionally three) juvenile stages (J1-J4), which, after four successive moulting events (J1-J2, J2-J3, J3-J4, J4-adult), result in the adult forms. The process often takes months to complete, but variations occur between different taxa (see Table 12.2). Every stage or step of the process requires a few weeks to be completed. Thus, the first moulting (J1-J2) of *D. stagnalis* takes place in 17-22 days, and the subsequent moults in 24-28, 30-35 and 46-52 days, respectively (Shafqat *et al.*, 1991). A marked increase in body volume is observed, so that, for example, the adult is 385 times larger than the embryo in the free-living form *Labronema ferox*, and 409 times in the plant-parasitic *Longidorus profundorum*. Nevertheless, the growth is visibly greater in smaller stages of the former and nearly constant in the latter (Yeates & Boag, 2002, 2003).

As mentioned above, the usual number of juvenile stages in dorylaims is four, but the existence of only three stages has been repeatedly reported in species of the genus *Xiphinema* (Halbrendt & Brown, 1992, 1993; Robbins *et al.*, 1996), species that do not seem to be closely related on morphological grounds. Such a compaction in the life cycle stages could be the result of occasional progenesis, a phenomenon not uncommon in parthenogenetic organisms (Coomans *et al.*, 2001) that often results in comparatively smaller adult forms (Yeates & Boag, 2002).

Leaving aside the changes affecting the caudal region, as already explained and discussed in a previous chapter, postnatal events involve major modifications of the cuticle (moulting) and genital system, and minor changes that modify other nematode organs and structures.

Moulting process

During moulting, the cuticle structures are replaced by another of a similar nature. It is associated with growth and seems necessary because extension of the cuticle is limited. When the diameter of the nematode increases, tangential stress also increases. The first signs of moulting are the release of the old cuticle from the underlying tissues and the hypertrophy of the latter. Then a new cuticle is secreted and finally ecdysis occurs when the old cuticle is broken (open) starting with a longitudinal slit that appears in the neck region. The details of this process in dorylaims has been well studied and narrated by several authors (Coomans & De Coninck, 1963; Coomans & van der Heiden, 1971; Grootaert & Lippens, 1974; Grootaert & Coomans, 1981; Carter & Wright, 1984; see also Wood, 1974; Shafqat *et al.*, 1991; Radivojević, 1998), whose main results are hereunder summarised.

The moulting of the anterior feeding apparatus consists of two events. Firstly, the formation of the cuticle lining, which includes the guiding sheath, the odontophore and the pharyngeal lining with its platelet structure. Secondly, the formation of the odontostyle. The cuticular lining is secreted by the muscle cells that partially differentiate during moulting and can be called myoepithelial cells: the Z-discs that were attached to the old cuticle disappear, the cells become glandular and many microtubules probably assist in the transport of the granules. After the secretion of a new cuticle the Z-discs re-appear. The cells forming part of the different sectors and subsectors of pharyngeal wall play an active role in this process.

Simultaneous with the formation of the new cuticle, a new replacement odontostyle is secreted at the base of the functional one. The formation process can be divided into two phases: *i*) phase A during which the odontostyle is secreted, with five subphases forming part of a continuous process, including three recognisable states, namely A_1: tip formed; A_3: odontostyle half-formed (Fig. 12.2A); A_5: odontostyle totally formed; and *ii*) phase B during which the replacement odontostyle migrates posteriorly into the pharyngeal wall while it contracts and hardens. During embryogenesis, immediately after the first (functional) odontostyle is formed, a second (replacement) one is secreted, preceding the formation of pharyngeal lining. The odontostyle is secreted inside the odontostyle forming cell, where the necessary substance is first synthesised in the cell body and deposited in a vacuole (phase A) and then condenses,

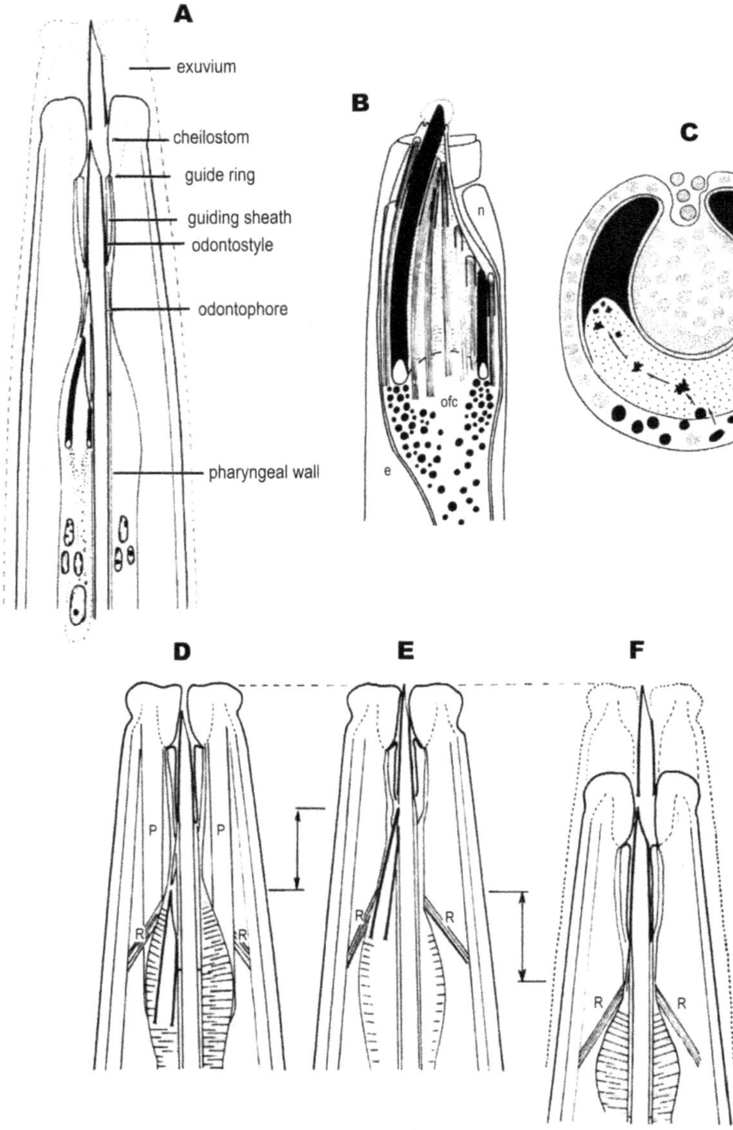

starts to contract and hardens (phase B) (Figure 12.2B, C). Regarding the migration, the replacement odontostyle of the J1 migrates anteriad a short distance, but in all other juvenile stages it migrates posteriad, probably because of a functional necessity to allow better movement of the anterior feeding apparatus. It has been demonstrated that the odontostyle twists on its axis when it moves forwards very quickly, but it is still unknown how it comes to occupy its correct position on the odontophore with the opening directed dorsally. The odontostyle also rotates during its posteriad migration into the odontostyle forming cell. Carter & Wright (1984) suggested that changes occurring in the shape of the two cells involved in odontostyle formation in *Xiphinema americanum* are largely responsible for movement of the odontostyle in the pharynx. The pharyngeal muscles, between which the odontostyle moves, are already functional but not yet completely attached. The tip of the replacement odontostyle may bear a 'protective cap' or other protection element around the tip to prevent piercing of the pharyngeal wall during the migration process.

The old odontostyle is expelled after several protrusions of the anterior feeding apparatus (Fig. 12.2D-F), whereas the old pharyngeal lining passes into the intestine. As the J1 bears a functional odontostyle when it hatches from the egg, and four odontostyles are formed throughout the postanatal development, a total of five odontostyles occur in the lifespan of each dorylaim. Exceptionally, however, the occurrence of a sixth, apparently non-functional odontostyle about 20% longer than the functional one has been observed in representatives of both free-living (Choudhary *et al.*, 2005; Ferris *et al.*, 2012) and plant-parasitic (Dalmasso, 1967; Loof & Maas, 1972; Zheng *et al.*, 2000) genera. The presence of a small odontostyle cap (mucron) inside the odontostyle forming cell has also been observed in adult specimens of several genera (summarised by Ferris *et al.*, 2012), and

Fig. 12.2. *Formation and migration of odontostyle during moulting events in dorylaims. A: Diagram of moulting in A_{2-3} stages (see text for additional information); B: Diagram of odontostyle formation (abbreviations: e = epidermis, n = nerve, ofc = odontostyle forming cell); C: Diagram of secretion and primary condensation of odontostyle (1 = membrane of vacuole, 2 = condensation of granular layer, 3 = condensation of stylet membrane, 4 = primary condensation of odontostyle); D-F: Diagram illustrating anteriad migration of replacement odontostyle during first moult (abbreviations: P = protractor muscles; R = retractor muscles.) (After Grootaert & Coomans (1981),* Nematologica *26(1980), pp. 419, 428, 430; Figs 2, 26, 27, 29.)*

tentatively interpreted as an incomplete cessation of the activity of that cell.

Genital system development

As explained by Shafqat *et al.* (1991) for *Dorylaimus stagnalis* (Figs 12.3, 12.4), the genital primordium of the J1 consists of two large granular germinal nuclei and 4-6 somatic nuclei while the ventral chord nuclei varies from 14 to 35 from the base of the pharynx to the middle of the primordium. The genital primordium of the J2 is made of 4-6 germinal and 8-14 somatic nuclei, and the ventral chord has 40-75 nuclei. The sexes can already be distinguished in the J3, in which the nuclei are 8-12, 20-45 and 170-190, respectively.

The juveniles that develop into females have 2-4 oval-shaped specialised ventral chord nuclei, and the middle part of the primordium bulges out slightly to form a cone-like structure representing the future vagina and vulva.

The juveniles that develop into males have a small compact mass of dark staining nuclei of spicule primordium posterior to the rectum, and a posteriorly directed tube (gonoduct) appears in the middle of the genital primordium, which forms the future genital duct.

The female J4 stage has 20-60 germinal nuclei in the primordium, whereas the number of somatic nuclei is very high. A few somatic nuclei are arranged on both sides of both the anterior and posterior arms along the wall of the developing gonad that form the epithelial covering of the gonad. The number of specialised ventral chord nuclei increases to 14-20 and then arrange themselves around the cone-like structure. With the completion of moulting the cuticle linings of the vagina and vulva are formed.

The male J4 stage has 18-61 germinal nuclei. The backwardly directed genital duct increases in size as the anterior and posterior arms of the developing system elongate. The somatic nuclei are arranged alternatively along the walls of the gonoduct. The spicule primordium becomes more compact and the spicules progressively appear. Many rows of darkly stained nuclei are arranged on the ventral side of the body between the cloaca and intestine-prerectum junction and give rise to copulatory muscles and the ventromedian supplements.

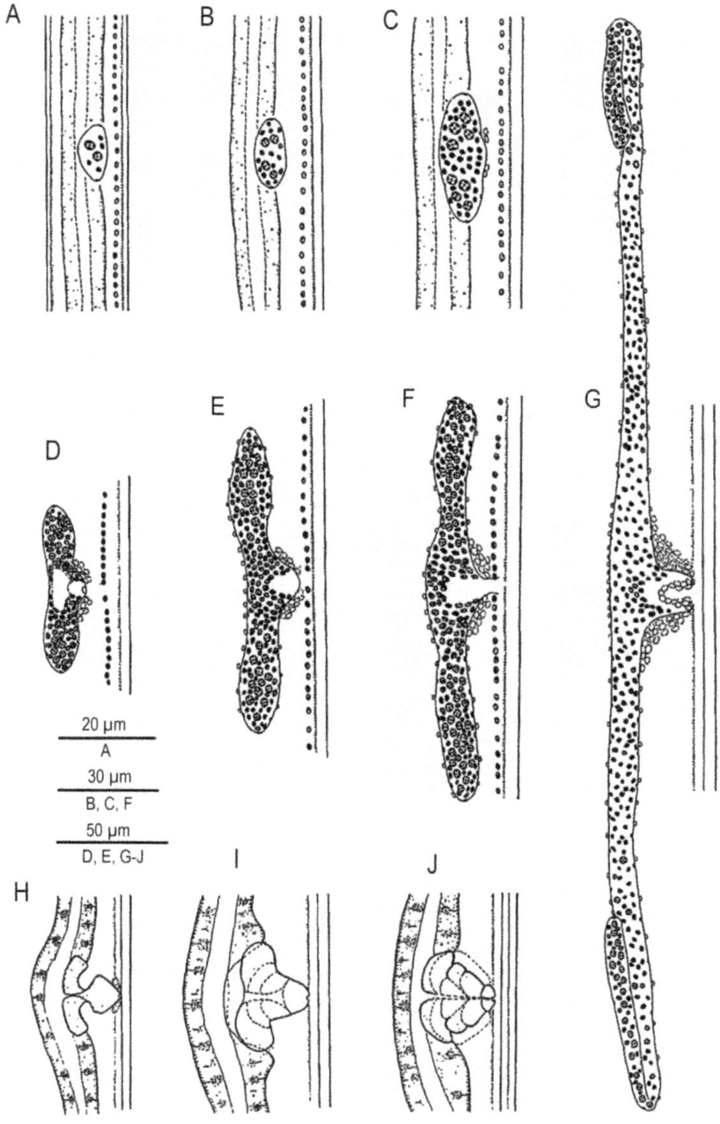

Fig. 12.3. *Postembryonic development of female genital system in* Dorylaimus stagnalis *(J1-J4 = corresponding juvenile stages from first to fourth). A: J1 primordium; B: J2 primordium; C: J3 primordium; D: J3 abnormal primordium; E: Primordium of moulting J3; F: J4 primordium; G: Primordium of moulting J4; H: Early J4; I: Mid-J4; J: Late J4. (After Shaftaq* et al. *(1991),* Revue de Nématologie *14, p. 67, Fig. 4.)*

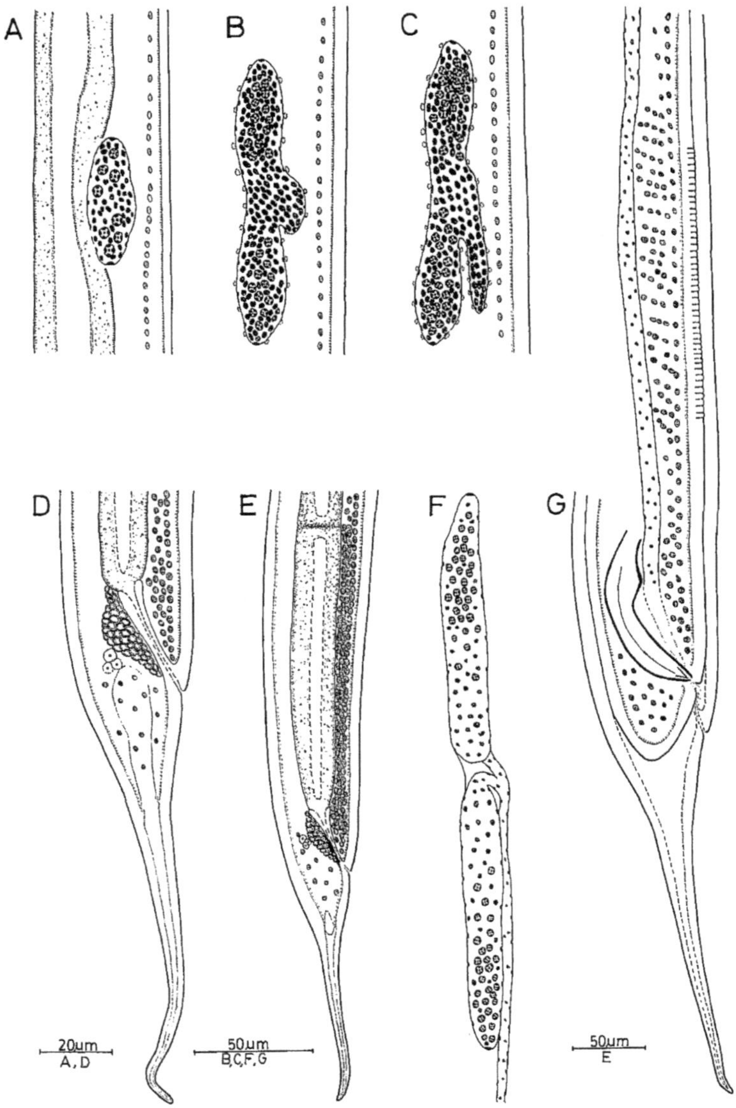

Fig. 12.4. *Postembryonic development of male genital system in* Dorylaimus stagnalis *(J3 and J4 = corresponding third- and fourth-stage juveniles). A: J3 primordium; B: Primordium of moulting J3; C: J4 primordium; D: Posterior body region of J3; E: Posterior body region of J4; F: Fully developed testes of moulting J4; G: Moulting J4; (After Shaftaq* et al. *(1991),* Revue de Nématologie *14, p. 68, Fig. 5.)*

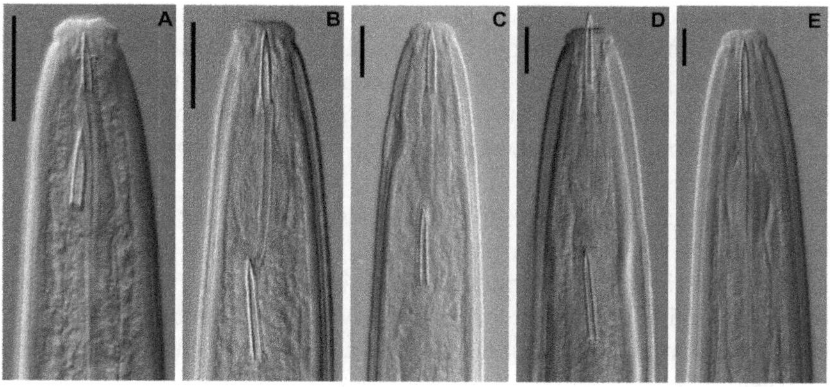

Fig. 12.5. *Postnatal development of* Nevadanema nevadense *showing the changes occurring in the anterior body region (J1-J4 = corresponding juvenile stages from first to fourth). A: J_1; B: J_2; C: J_3; D: J_4; E: Adult female. (After Álvarez-Ortega & Peña-Santiago (2013a),* Journal of Nematode Morphology and Systematics *16, p. 169, Fig. 1A-E.) (Scale bars = 20 μm.)*

Samsoen & Barbez (1982) studied the development of the genital system in *Xiphinema* species and found that the development of the genital primordium is slow in the first three juvenile stages, with no relevant differences between the stages, but then enlarges more in the J4 as a result of several cell divisions, drastic changes only occurring during the last moult. The same authors pointed out that the retardation in growth of the genital primordium seemed to be related to the formation of the odontostyle, a process that may be exigent and does not allow a large change in the genital primordium.

Other (minor) postnatal changes

The lip region is liable to undergo small but perceptible changes. Thus, Álvarez-Ortega & Peña-Santiago (2013a) found that the lip region in *Nevadanema nevadense* starts off being quite angular in the J1 and J2 stages (Fig. 12.5A, B, respectively), and becomes more rounded in the J3 and J4 stages (Fig. 12.5C, D, respectively) and in adult specimens (Fig. 12.5E).

The appearance of the odontostyle may be significantly modified during ontogeny. For example, the J1 of *N. nevadense* bears a robust odontostyle (Fig. 12.5A), nearly as long as the lip region diameter and with a distinct aperture, whereas subsequent juvenile stages (Fig. 12.5B-E) display a more slender odontostyle that is longer than the lip region diameter and has a progressively shorter aperture, the changes observed being statistically significant when the corresponding morphometrics are compared. The nature of the guiding ring also may be modified, from a simple to a double type.

The structure of pharynx may affect or influence the postembryonic development of this organ (*cf.* Geraert, 1978). Thus, the taxa whose pharynx consists of a non-muscular anterior part and a basal bulb result in a comparatively shorter adult pharynx (with less postnatal growth rate) than those forms displaying a muscular anterior part and a larger basal expansion.

The conical cardia of *Dorylaimus stagnalis* is made up of tissues in the adults, but of cardiac cells in the juveniles. At each moulting stage, and with the formation of the odontostyle and other body parts, a new cardia is formed by multiplication of cardiac cells. The number of these cells is variable: 5-8, 9-12, 11-14 and 13-16 in J1, J2, J3 and J4, respectively (Shafqat *et al.*, 1991).

Life cycle and life span

Dorylaims are K-strategists, with a comparatively long life span (Table 12.2). Thus, their embryonic development may take 10-12 days in *Aporcelaimellus* to nearly 1 month in some longidorid species. The postembryonic development (from hatching to adult stage) is often completed in a few months, but extending for at least 2 years in some cases. Thus, it takes 117-138 days in females and 111-131 days in males of *Dorylaimus stagnalis*, but 2 or 3 years in longidorids. Life span extends for up to 6 years in some *Xiphinema* species.

References

ÁLVAREZ-ORTEGA, S. & PEÑA-SANTIAGO, R. (2013a). The postembryonic development of *Nevadanema nevadense* (Dorylaimida, Qudsianematidae) reveals more relevant changes than expected. *Journal of Nematode Morphology and Systematics* 16, 167-172.

ÁLVAREZ-ORTEGA, S. & PEÑA-SANTIAGO, R. (2013b). Taxonomy of the genus *Aporcelaimellus* Heyns, 1965 (Nematoda, Dorylaimida, Aporcelaimidae). *Zootaxa* 3669, 243-260. DOI: 10.11646/zootaxa.3669.3.3

ÁLVAREZ-ORTEGA, S. & PEÑA-SANTIAGO, R. (2014). Redescription of *Pungentus pungens* Thorne & Swanger, 1936 (Dorylaimida: Nordiidae), with a revised taxonomy of the genus. *Nematology* 16, 387-401. DOI: 10.1163/15685411-00002766

ÁLVAREZ-ORTEGA, S. & PEÑA-SANTIAGO, R. (2019). Morphology, phylogeny and taxonomy of the genus *Sectonema* (Nematoda, Aporcelaimidae). *Zoologica Scripta* 48, 535-544. DOI: 10.1111/zsc.12357

ÁLVAREZ-ORTEGA, S., SUBBOTIN, S.A. & PEÑA-SANTIAGO, R. (2013a). Morphological and molecular characterisation of *Aporcelaimellus simplex* (Thorne & Swanger, 1936) Loof & Coomans, 1970 and a new concept for *Aporcella* Andrássy, 2002 (Dorylaimida: Aporcelaimidae). *Nematology* 15, 165-178. DOI: 10.1163/156854112X651320

ÁLVAREZ-ORTEGA, S., SUBBOTIN, S.A. & PEÑA-SANTIAGO, R. (2013b). Morphological and molecular characterisation of Californian species of *Metaporcelaimus* Lordello, 1965 (Dorylaimida: Aporcelaimidae), with a new concept of the genus. *Nematology* 15, 251-278. DOI: 10.1163/15685411-00002674

ÁLVAREZ-ORTEGA, S., NGUYEN, T.A.D., ABOLAFIA, J., VU, T.T.T. & PEÑA-SANTIAGO, R. (2015). Three new species of the genus *Aporcelaimoides* Heyns, 1965 from Vietnam (Nematoda, Dorylaimida, Aporcelaimidae), with an updated taxonomy of the genus. *Zookeys* 516, 1-26. DOI: 10.3897/zookeys.516.10087

ANDRÁSSY, I. (2009). *Free-living nematodes of Hungary, III (Nematoda errantia)*. In: *Pedozoologica Hungarica 5* (Series Editors: Csuzdi, C. & Mahunka, S.). Budapest, Hungary, Hungarian Natural History Museum and Systematic Zoology Research Group of the Hungarian Academy of Sciences.

ASGARI, M., ESKANDARI, A., KARANI, H.M., ABOLAFIA, J. & PEÑA-SANTIAGO, R. (2020). Morphological and molecular characterization of *Lenonchium zanjanense* sp. n. (Dorylaimida, Nordiidae) from Iran, with new insights into the phylogeny of the genus and a revision of its taxonomy. *Journal of Helminthology* 94, e47, 11 pp. DOI: 10.1017/S0022149X19000233

BOAG, B. (1985). Effect of temperature on the times to hatching of eggs of the plant parasitic nematode *Longidorus elongatus*. *Nematologia Mediterranea* 13, 61-66.

BROWN, D.J.F. & COIRO, M.I. (1983). The total reproductive capacity and longevity of individual female *Xiphinema diversicaudatum* (Nematoda: Dorylaimida). *Nematologia Mediterranea* 11, 87-92.

BROWN, D.J.F. & COIRO, M.I. (1985). The reproductive capacity and longevity of *Xiphinema index* (Nematoda: Dorylaimida) from three populations on selected host plants. *Revue de Nématologie* 8, 171-173.

CARTER, R.F. & WRIGHT, K.A. (1984). Formation of the odontostyle during moulting of the nematode *Xiphinema americanum* (Nematoda: Dorylaimoidea). *Journal of Ultrastructure Research* 87(1983), 221-241. DOI: 10.1016/S0022-5320(84)80062-3

CHOUDHARY, M., AHMAD, W., TAHSEEN, Q. & JAIRAJPURI, M.S. (2005). The occurrence of sixth odontostyle in *Dorylaimoides* sp. *Journal of Nematode Morphology and Systematics* 7(2004), 169-172.

COHN, E. & MORDECHAI, M. (1969). Investigations on the life cycle and host preference of some species of *Xiphinema* and *Longidorus* under controlled conditions. *Nematologica* 15, 295-302. DOI: 10.1163/187529269X00335

COIRO, M.I. & AGOSTINELLI, A. (1991). The development of juvenile stages of *Xiphinema index* (Nematoda: Dorylaimida) on *Vitis vinifera*. *Revue de Nématologie* 14, 181-182.

COIRO, M.I. & SASANELLI, N. (1995). Life-cycle studies of individual *Longidorus athesinus* (Nematoda) on S. Lucie cherry. *Nematologia Mediterranea* 23, 329-333.

COIRO, M.I., SERINO, M. & AGOSTINELLI, A. (1991). Feeding and reproduction of *Xiphinema index* (Nematoda: Dorylaimida) on two hosts at three temperatures. *Nematologia Mediterranea* 19, 101-102.

COIRO, M.I., SASANELLI, N. & SERINO, M. (1995). Fecundity and longevity of individual *Xiphinema ifacolum* (Nematoda: Dorylaimidae) on tomato. *Nematologica* 41, 191-196. DOI: 10.1163/003925995X00161

COIRO, M.I., LAMBERTI, F., SABOVÁ, M., AGOSTINELLI, A., VALOCKÁ, B. & SASANELLI, N. (1999). The life cycle of a population of *Xiphinema diversicaudatum* from Slovakia (Nematoda). *Nematologia Mediterranea* 28, 27-30.

COIRO, M.I., LAMBERTI, F., SABOVÁ, M., SASANELLI, N., AGOSTINELLI, A. & VALOCKÁ, B. (2000). The life cycle of *Xiphinema vuittenezi* (Nematoda) from Slovakia. *Nematologia Mediterranea* 28, 27-30.

COIRO, M.I., LAMBERTI, F., SASANELLI, N., DUNCAN, L.W. & AGOSTINELLI, A. (2002). Reproduction and longevity of *Xiphinema vulgare* (Nematoda). *Nematologia Mediterranea* 30, 91-95.

COOMANS, A. & DE CONINCK, L.A.P. (1963). Observations on spear formation in *Xiphinema*. *Nematologica* 9, 85-96. DOI: 10.1163/187529263X00179

COOMANS, A. & VAN DER HEIDEN, A. (1971). Structure and formation of the feeding apparatus in *Aporcelaimus* and *Aporcelaimellus* (Nematoda: Dorylaimidae). *Zeitschrift für Morphologie und Ökologie der Tiere* 70, 103-118.

COOMANS, A., HUYS, R., HEYNS, J. & LUC, M. (2001). Character analysis, phylogeny and biogeography of the genus *Xiphinema* Cobb, 1913 (Nematoda: Longidoridae). *Annales Sciences Zoologiques, Musée Royal de l'Afrique Centrale, Teureven, Belgique* 287, 1-239.

DALMASSO, A. (1967). Cas de femelles de *Xiphinema* possédant deux odontostyles. *Nematologica* 13, 474-476. DOI: 10.1163/187529267X00788

DALMASSO, A. (1975). Cytogenetics and reproduction in *Xiphinema* and *Longidorus*. In: Lamberti, F., Taylor, C.E. & Seinhorst, J.W. (Eds). *Nematode vectors of plant viruses*. London and New York, Plenum Press and NATO Scientific Affairs Division, pp. 139-151.

FERRIS, H., ROBBINS, R. & YEATES, G. (2012). Atypical development in plant and soil nematodes. *Journal of Nematology* 44, 1-6.

FERRIS, V.R. (1968). Biometric analyses in the genus *Labronema* (Nematoda: Dorylaimida) with description of *L. thornei* sp. n. *Nematologica* 14, 276-284. DOI: 10.1163/187529268X00480

FLEGG, J.J.M. (1966). Once-yearly reproduction in *Xiphinema vuittenezi*. *Nature* 212, 741. DOI: 10.1038/212741a0

FLEGG, J.J.M. (1968a). Genital tract development in *Xiphinema vuittenezi*. *Nematologica* 14, 640-641. DOI: 10.1163/187529267X00472

FLEGG, J.J.M. (1968b). Life-cycle studies of some *Xiphinema* and *Longidorus* species in southeastern England. *Nematologica* 14, 197-210. DOI: 10.1163/187529268X00426

FLEGG, J.J.M. (1968c). Embryonic studies of some *Xiphinema* and *Longidorus* species. *Nematologica* 14, 137-145. DOI: 10.1163/187529268X00732

GERAERT, E. (1978). On growth and form in nematodes: II. Oesophagus and body width in Dorylaimida. *Nematologica* 24, 347-360. DOI: 10.1163/187529278X00470

GILARTE PADILLA, P., CARREIRA, J.A. & PEÑA-SANTIAGO, R. (2013). Description of *Paravulvus moroccanus* sp. n. (Nematoda, Dorylaimida, Nygolaimidae) from the High Atlas Mountains, Morocco, with notes on the taxonomy of the genus. *Zootaxa* 3734, 545-558. DOI: 10.11646/zootaxa.3734.5.4

GROOTAERT, P. & COOMANS, A. (1981). The formation of the anterior feeding apparatus in dorylaims. *Nematologica* 26(1980), 406-431. DOI: 10.1163/187529280X00350

GROOTAERT, P. & LIPPENS, P.L. (1974). Some ultrastructural changes in cuticle and hypodermis of *Aporcelaimellus* during the first moult (Nematoda: Dorylaimoidea). *Zeitschrift für Morphologie und Ökologie der Tiere* 79, 269-282.

GROOTAERT, P. & SMALL, R.W. (1982). Aspects of the biology of *Labronema vulvapapillatum* (Meyl) (Nematoda: Dorylaimida) in laboratory culture. *Biologisch Jaarboek Dodonaea* 50, 135-148.

HALBRENDT, J.M. & BROWN, D.J.F. (1992). Morphometric evidence for three juvenile stages in some species of *Xiphinema americanum* sensu lato. *Journal of Nematology* 24, 305-309.

HALBRENDT, J.M. & BROWN, D.J.F. (1993). Aspects of biology and development of *Xiphinema americanum* and related species. *Journal of Nematology* 25, 355-360.

HARRIS, A.R. (1979). Seasonal populations of *Xiphinema index* in vineyard soils of north-eastern Victoria, Australia. *Nematologica* 25, 336-347. DOI: 10.1163/187529279X00073

HOANG, H., CHU, H.H., NGUYEN, T.A.D., TRINH, Q.P., ABOLAFIA, J. & PEÑA-SANTIAGO, R. (2019). Morphological and molecular characterisation of *Chitwoodius coffeae* sp. n. (Dorylaimida; Tylencholaimidae) from Vietnam, with a revised taxonomy of the genus. *Nematology* 21, 509-521. DOI: 10.1163/15685411-00003231

HUNT, D.J. (1978). Bionomics of *Paractinolaimus vigor* Thorne, 1967 (Dorylaimida: Paractinolaimidae) with a description of *P. dominicus* n. sp. and *Nygolaimium haguei* n. sp. (Dorylaimida: Aporcelaimidae). *Nematologica* 23(1977), 452-462. DOI: 10.1163/187529277X00381

JAFFEE, B.A., HARRISON, M.B., SHAFFER, R.L. & STRANG, M.B. (1987). Seasonal population fluctuation of *Xiphinema americanum* and *X. rivesi* in New York and Pennsylvania orchards. *Journal of Nematology* 19, 369-378.

JAIRAJPURI, M.S. & BAJAJ, H.K. (1978). Observations on the biology of *Xiphinema basiri* and *X. insigne*. *Revue de Nématologie* 1, 227-239.

JATALA, P. (1975). Endotokia matricida in a *Xiphinema* sp. *Journal of Nematology* 7, 205-206.

KOLODGE, C., RADEWALD, J.D. & SHIBUYA, F. (1987). Revised host range and studies on the life cycle of *Longidorus africanus*. *Journal of Nematology* 19, 77-81.

KRUGER, J.C. DE W. & HEYNS, J. (1988). Notes on the cytogenetics of some South African *Xiphinema* species (Nematoda: Longidoridae). *South African Journal of Zoology* 23, 337-341.

LOOF, P.A.A. & MAAS, P.W.T. (1972). The genus *Xiphinema* (Dorylaimida) in Surinam. *Nematologica* 18, 92-119. DOI: 10.1163/187529272X00287

PEÑA-SANTIAGO, R. & VARELA, I. (2017). Taxonomy and systematics of the genus *Makatinus* Heyns, 1965 (Nematoda: Dorylaimida: Aporcelaimidae). *Journal of Nematology* 49, 245-253. DOI: 10.21307/jofnem-2017-070

PEÑA-SANTIAGO, R., ABOLAFIA, J. & PEDRAM, M. (2012). New data on *Labronemella labiata* Andrássy, 1985 (Dorylaimida: Qudsianematidae) from Iran, with SEM study and a compendium of the genus. *Zootaxa* 3271, 43-54. DOI: 10.11646/ZOOTAXA.3271.1.4

PEÑA-SANTIAGO, R., NIKNAM, G., ÁLVAREZ-ORTEGA, S. & JABBARI, H. (2014a). *Metaxonchium persicum* sp. n. from Iran (Nematoda, Dorylaimida, Belondiridae), with an updated taxonomy of the genus. *Zootaxa* 3785, 501-517. DOI: 10.11646/zootaxa.3785.4.1

PEÑA-SANTIAGO, R., LIÉBANAS, G. & AHMAD, W. (2014b). Taxonomy of the genus *Oriverutus* Siddiqi, 1971 (Nematoda: Dorylaimida: Nordiidae). *Journal of Natural History* 48, 1253-1278. DOI: 10.1080/00222933.2013.861942

PEÑA-SANTIAGO, R., GUERRERO, P., LIÉBANAS, G., GARCÍA, M.C., PALOMEQUE, T. & LORITE, P. (2015). Characterisation of an Iberian population of *Rhyssocolpus iuventutis* Andrássy, 1971 (Dorylaimida: Nordiidae), with a revised taxonomy of the genus. *Nematology* 17, 139-153. DOI: 10.1163/15685411-00002857

PEÑA-SANTIAGO, R., ABOLAFIA, J., NGUYEN, T.A.D., ÁLVAREZ-ORTEGA, S. & VARELA-BENAVIDES, I. (2019). General morphology and updated taxonomy of the genus *Aporcelinus* Andrássy, 2009 (Dorylaimida: Aporcelaimidae). *Nematology* 21, 857-874. DOI: 10.1163/15685411-00003260

PROTA, U. & GARAU, R. (1973). Indagini sulla biologia di *Xiphinema index* Thorne & Allen in vigneti sardi. *Nematologia Mediterranea* 1, 36-54.

PROTA, U., BLEVE-ZACHEO, M.T., GARAU, R. & LAMBERTI, F. (1977). Studies on the life-cycle of *Xiphinema index*. *Nematologia Mediterranea* 5, 299-303.

RADIVOJEVIĆ, M. (1998). Observations on spear replacement in *Xiphinema dentatum* (Nematoda: Longidoridae). *Nematologica* 44, 137-151. DOI: 10.1163/005325998X00036

RASCHKÉ, I.E. & BOAG, B. (1981). Morphological variation and distribution frequency of hermaphrodite *Longidorus elongatus*. *Revue de Nématologie* 4, 283-285.

ROBBINS, R.T., BROWN, D.J.F., HALBRENDT, J.M. & VRAIN, T.C. (1996). Compendium of juvenile stages of *Xiphinema* species (Nematoda: Longidoridae). *Russian Journal of Nematology* 4, 163-171.

SAMSOEN, L. & BARBEZ, D. (1982). The development of the genital system in *Xiphinema* and the use of genital primordial for separating juvenile stages. *Medeelinge Faculteit Landbouwhogeschool Rijksuniversiteit Gent* 47, 771-780.

SHAFQAT, S., JAIRAJPURI, M.S. & BILGRAMI, A.L. (1991). Developmental biology of *Dorylaimus stagnalis* Dujardin, 1845 (Nematoda: Dorylaimida). *Revue de Nématologie* 14, 61-71.

SHAFQAT, S., AHMAD, I. & AHMAD, W. (1994). SEM morphology and embryology of *Laevides symmetricus* (Nygolaimina: Nygolaimidae). *Nematologica* 40, 534-540. DOI: 10.1163/003525994X00409

SHAFQAT, S., AHMAD, I. & JAIRAJPURI, M.S. (1996). Early embryonic development patterns in some predacious dorylaims. *Afro-Asian Journal of Nematology* 6, 133-136.

THOMAS, P.R. (1969). Population development of *Longidorus elongatus* on strawberry in Scotland with observations on *Xiphinema diversicaudatum* on raspberry. *Nematologica* 15, 582-590. DOI: 10.1163/187529269X00902

TRUDGILL, D.L. (1976). Observations on the mating of *Xiphinema diversicaudatum*. *Nematologica* 22, 227. DOI: 10.1163/187529276X00355

VOVLAS, V. & LARIZZA, A. (1994). Life stages identification and embryology of the virus-vector nematode *Xiphinema index*. *Nematologia Mediterranea* 22, 109-113.

WOOD, F.H. (1973). Nematode feeding relationships. Feeding relationships of soil-dwelling nematodes. *Soil Biology and Biochemistry* 5, 528-537. DOI: 10.1016/0038-0717(73)90049-7

WOOD, F.H. (1974). Biology of *Aporcelaimellus* sp. (Nematoda: Aporcelaimidae). *Nematologica* 19 (1973), 528-537. DOI: 10.1163/187529273X00547

WYSS, U. (1970). Untersuchungen zur Populationsdynamik von *Longidorus elongatus*. *Nematologica* 16, 74-84. DOI: 10.1163/187529270X00496

YASSIN, A.M. (1969). Glass-house and laboratory studies on the biology of the needle nematode, *Longidorus elongatus*. *Nematologica* 15, 169-178. DOI: 10.1163/187529269X00597

YEATES, G.W. & BOAG, B. (2002). Post-embryonic growth of longidorid nematodes. *Nematology* 4, 883-889. DOI: 10.1163/156854102321122494

YEATES, G.W. & BOAG, B. (2003). Growth and life histories in Nematoda with particular reference to environmental factors. *Nematology* 5, 653-664. DOI: 10.1163/156854103322746832

ZAHEDI, E., NIKNAM, G., JABBARI, H. & PEÑA-SANTIAGO, R. (2016). Two new species of the rare genus *Nygolaimoides* Meyl *in* Andrássy, 1960 (Nematoda, Dorylaimida, Thorniidae) from Iran, with a compendium of its species. *Zootaxa* 4150, 61-72. DOI: 10.11646/zootaxa.4150.1.3

ZHENG, J., PENEVA, V. & BROWN, D.J.F. (2000). *Longidorus camelliae* n. sp. (Nematoda: Longidoridae) associated with ornamental cultivars of *Camellia japonica* L. growing in a nursery at Fuyang, Zheijang Province, eastern China. *Systematic Parasitology* 47, 119-125. DOI: 10.1023/A:1006482322716

Nematology Monographs & Perspectives, 2021, Vol. 13, 193-229

Chapter 13

Ecology and biogeography

Rationale – dorylaims as K-strategists

Dorylaims are major components of the nematode communities dwelling in continental habitats, either soils or freshwater sediments. They make an interesting group for environmental studies, not only because of their number and diversity, but also their comparatively large size and biomass, their omnivorous feeding habits, and their traits as K-strategists. Unfortunately, the ecology of the group, especially that of free-living forms, has received little attention and many aspects remain unknown (*cf.* McSorley, 2012).

The proposal of maturity indices (Bongers, 1990; Bongers *et al.*, 1995; Bongers & Ferris, 1999; Ferris & Bongers, 2009) opened a fruitful 30-year period of progress in nematode ecology, with interesting and significant results in several research lines: bioindicators, the role of nematodes in soil food webs, the elucidation of structural and functional aspects of nematode communities, and others. Nonetheless, most contributions put emphasis on nematode assemblages (trophic groups, guilds, *etc.*), whereas the study of particular species did not generate much interest. One of the main impediments to decipher the relationships between nematode species and environmental variables is the difficulty inherent in identifying the taxa with accuracy, a serious handicap in the case of dorylaims. As a result, perhaps, many ecological works on dorylaims dealt with either genera (McSorley, 2012) or families (Coll *et al.*, 2012), and were much less often focused on specific taxa (Liébanas *et al.*, 2002, 2004), despite the fact that several authors (Yeates, 2003; Háněl, 2010) have claimed that species-level identification is an essential requirement to make relevant progress in this field.

The ecological role of dorylaims in the soils and sediments they inhabit is not yet well elucidated due to insufficient information about their feeding habits and life strategies. Available evidence, however, supports them as being K-strategists with low generation time, low reproduction rates, low metabolic activity, and slow movement. In fact, dorylaims are assigned values 4 and 5 in the coloniser-persister series (ranging from 1 to 5) originally proposed by Bongers and others for calculating maturity indices. In addition, their comparatively large size (biomass) compared to other soil-dwelling nematodes results in a greater impact on energy flow and makes for a larger metabolic footprint (Ferris, 2010; McSorley, 2012).

Some observational and experimental studies have confirmed the persister features of dorylaims. For example, both primary and secondary nematode succession in artificial (residue decomposition in plots), semi-natural (fallow) and natural habitats (pine forests, *etc.*) show that most dorylaimid forms, namely those belonging to the genera *Aporcelaimellus*, *Crassolabium*, *Ecumenicus*, *Eudorylaimus* and *Mesodorylaimus*, among others, occur in later or advanced steps of the process (Háněl, 2001, 2003, 2010; Georgieva *et al.*, 2005). Nevertheless, when species identification is refined, the results obtained are more difficult to interpret since, as discussed by Háněl (2010: p. 368) regarding the case of a few members of *Aporcelaimellus*, "...one genus, which evidently is omnivorous, K-strategic ..., can include species that can either behave as colonisers or persisters. Similar behavioural differences may also occur within other omnivorous dorylaimid genera and may complicate the application of functional guilds in soil nematode studies".

Dorylaims as components of nematode communities

The ecological role played by dorylaims in their communities is certainly far from being well elucidated, but the analysis of some fundamental parameters (diversity, abundance and biomass) can be useful to provide a quantitative approximation to the matter.

Dorylaimid diversity, expressed as the number of species collected in a particular habitat or a sampling unit, often termed species richness, is extremely variable and clearly depends on environmental factors. Available data suggest that when ecological studies are implemented after a detailed taxonomic study, with reliable identification to species level, dorylaimid forms very often represent at least one-fifth of the total nematode species (Table 13.1). This ratio apparently occurs in both natural habitats and agrosystems, in terrestrial soils and freshwater

Table 13.1. *Nematode (total) and dorylaimid species richness*[1] *in different kind of habitats.*

Habitat	Location	S_N^2	S_D^3	%	Country	Reference
Dune sand	1	31	8	25.81	New Zealand	Yeates (1967)
	2	32	9	28.13		
	3	24	7	29.17		
	4	20	3	15.00		
	5	32	4	12.50		
	6	22	5	22.73		
	Average	26.83	6.00	22.22		
Meadow	1	98	24	24.49	Romania	Popovici (1998)
	2	83	18	21.69		
	3	76	18	23.68		
	4	52	12	23.08		
	5	56	10	17.86		
	6	50	10	20.00		
	7	64	10	15.87		
	Average	68.3	14.6	20.95		
Beech forest	1	65	12	18.46	Romania	Popovici (1995)
	2	66	10	15.15		
	3	66	12	18.18		
	4	77	16	20.78		
	5	84	14	16.67		
	6	74	15	20.27		
	7	64	11	17.19		
	Average	70.8	12.9	18.10		
Maize crop	1	22	6	27.27	Czech Republic	Cerevková & Cagán (2015)
	2	27	6	22.22		
	3	26	7	26.92		
	4	34	10	29.41		
	Average	27.25	7.25	26.46		
Freshwater	1	10	3	30.0	Croatia	Vidaković & Bogut (2006)
	2	16	7	43.7		
	3	10	4	40.0		
	Average	12	4.7	37.9		

[1] Data calculated from the original information provided in the corresponding reference.
[2] S_N = total nematode species richness.
[3] S_D = dorylaimid species richness.

sediments, and it is also noted in different successional stages of ecosystems and distant geographical areas.

The number of specimens (abundance) that make up the nematode assemblage is another relevant variable to establish its structure, presenting much higher (absolute) values than those of species richness (Table 13.2). The relative abundance of dorylaims, *i.e.*, their percentage of the total nematode abundance, is lower and more variable than that observed for species richness. Thus, on the one hand, dorylaims in general present lower values than those observed for species richness, which means that each dorylaimid species is represented by a comparatively lower number of specimens in the assemblage. On the other hand, and when data of different habitats are compared, the maximum value is four times the minimum one; meanwhile the maximum value is only twice the minimum in the case of species richness (Table 13.1).

The general large size of dorylaims compared to other nematode taxa results in high values for their biomass. Actually, this parameter probably represents the most remarkable contribution of the group to nematode community, much exceeding their influence in terms of diversity and abundance. Thus, Zullini *et al.* (2011) provided biomass data of 90 nematode species, including 23 dorylaimid, dwelling in freshwater springs in Italy. The biomass of dorylaimid species was 6.41 ± 7.67 (0.22-29.72) μg per adult specimen, whereas the corresponding values for the remaining 67 non-dorylaimid species were 1.90 ± 3.03 (0.02-15.94) μg. Notably, seven out the ten species having the highest values of biomass were dorylaims (Table 13.3).

Local distribution

Distribution of dorylaims at a small (local) scale, referring to their occurrence (presence/absence), frequency and abundance in single sites or reduced areas, is essentially determined by microclimate, soil properties and vegetation (Weischer & Almeida, 1995). Nevertheless, in general, it may be affected by many environmental variables. The available information concerning the effect of these variables on the distribution of a particular species is very rich, but the results do not always match. Some examples that illustrate the complexity of the subject are presented below.

Table 13.2. *Nematode (total) and dorylaimid abundance[1] in different kind of habitats.*

Habitat	Location	Nn[2]	Nd[3]	%	Country	Reference
Sand dunes	1	427	99	23.2	New Zealand	Yeates (1967)[4]
	2	2082	238	11.4		
	3	597	325	54.4		
	4	1811	411	22.7		
	5	482	109	22.6		
	6	67	14	20.9		
	Average	911	199	25.9		
Meadow	1	975	243	24.9	Romania	Popovici (1998)[5]
	2	798	107	13.4		
	3	590	34	5.7		
	4	1445	145	10.0		
	5	3125	444	14.2		
	6	3252	250	7.7		
	7	1720	200	11.6		
	Average	1700	203	12.5		
Beech forest	1	921	120	13.0	Denmark	Yeates (1973)[5]
Maize crop	1	357	40	11.3	Czech	Cerevková &
	2	183	39	21.3	Republic	Cagán (2015)[6]
	3	465	78	16.8		
	4	172	30	17.6		
	Average	273	47	16.5		
Freshwater	1	233	19	8.1	Croatia	Vidaković &
	2	556	31	5.6		Bogut (2006)[7]
	3	174	11	6.3		
	Average	321	20.3	6.7		

[1] Data calculated from the original information provided in the corresponding reference.
[2] N_n = Total number of nematode specimens.
[3] S_D = Number of dorylaimid specimens.
[4] Number of specimens (1000 g soil)$^{-1}$.
[5] Number of specimens ($\times$ 10^3) m^{-2}.
[6] Number of specimens (950 g soil)$^{-1}$.
[7] Number of specimens (100 g macrophyte d wt)$^{-1}$.

Many studies reveal the influence of soil properties on the distribution of dorylaims. Although it has been stated that soil texture probably does not impose much restriction (McSorley, 2012), it is reasonable

Table 13.3. *Ten nematode species having the highest values of biomass out of a total of 90 species inhabiting freshwater springs in Italy. Biomass values given as average weight in μg per adult specimen. After Zullini* et al. *(2011).*

Species	Biomass (μg)
Dorylaimus stagnalis	29.72
Aporcelaimellus krygeri	22.14
Aporcelaimellus obtusicaudatus	16.25
Ironus paramacramphis	15.94
Prodorylaimus longicaudatoides	14.48
Ironus tenuicaudatus	10.58
Paraxonchium cf. *laetificans*	10.33
Tripyla glomerans	10.16
Aporcelaimellus sp. 1	9.45
Paractinolaimus macrolaimus	9.19

to assume that the size of soil particles may influence the movement of selected species depending on their dimensions. Thus, Norton *et al.* (1971) observed a negative correlation of *Xiphinema americanum* with the clay content, and Liébanas *et al.* (2004) found that clay content may explain the distribution of several species with significant (either positive or negative) correlation. Chemical composition of soils probably influences the biology of many species. For example, it has been demonstrated that low levels of pH result in a decrease or total absence of some dorylaimid genera (Norton & Hoffmann, 1974; McSorley, 2012), and in an increase of *Xiphinema americanum* (Norton *et al.*, 1971; Niblack & Bernard, 1985) and several other species (Liébanas *et al.* (2004). De Waele & Coomans (1990) also found different effects of this variable on the distribution of longidorid nematodes in Belgium. The level of some essential soil elements can play an important role in the distribution of dorylaimid species, as observed by Liébanas *et al.* (2004) who found that two species of the genus *Crassolabium*, namely *C. eroshenkoi* and *C. ettersbergense*, show positive and negative correlation, respectively, with the nitrogen content of soil. The presence of some soil components (cations, salts, heavy metals) often results in a negative effect on dorylaimid populations (Norton *et al.*, 1971), but this effect may differ depending on metal concentration and the dorylaimid species considered (for a discussion see McSorley, 2012). Some physical soil properties have been less analysed, but Weischer (1975) and Nath *et al.* (1998) provided examples of the different behaviour of longidorid species regarding soil temperature, whereas Liébanas *et al.* (1994) observed that soil colour (measured as 'colour chroma') may partially explain the distribution of *Aporcelaimellus* sp., *Xiphinema pachtaicum* and *Heterodorus brevidentatus*, although not necessarily with the same effect.

Taking into account that, like other nematodes, dorylaims are basically aquatic organisms, soil moisture might condition their biology due to an increase in survival, reproduction rates, hatching, emergence from anhydrobiosis, *etc.* In fact, Porazinska *et al.* (1998) noted that the total numbers and relative proportions of *Aporcelaimellus* and *Eudorylaimus* populations steadily increased with irrigation, but other authors (Weischer, 1975; McSorley, 2012) pointed out that watering may have a positive or neutral effect on dorylaimid populations. Regarding rainfall, McSorley (1997) found significant differences in the densities of *Aporcelaimellus*, *Eudorylaimus* and *Mesodorylaimus* populations inhabiting a citrus culture and an adjacent pasture in correlation with this variable, and Nath *et al.* (1998) obtained similar results in the seasonal distribution of *Xiphinema radicicola*.

Management of soils often results in changes to their properties and functionality, which influences the distribution of dorylaims. For example, chemical, organic and compost fertilisation is known to produce different (positive or negative) effects on the populations of more than a dozen dorylaimid genera (Fiscus & Neher, 2002; Ferris & Matute, 2003; Hu & Qi, 2010; Zhao & Neher, 2013). McSorley & Frederick (1999) found that organic fertilisation produces significant differences, often representing population increases, of several dorylaimid genera, either fungivorous (*Leptonchus* and *Tylencholaimus*), omnivorous (*Ecumenicus*, *Eudorylaimus* and *Mesodorylaimus*) or predators (*Discolaimus* and *Nygolaimus*). Tillage or farming techniques also modify the abundance (number of individuals) of several genera (Porazinska *et al.*, 1999; Fiscus & Neher, 2002; McSorley, 2012, 2013; Postma-Blaauw *et al.*, 2012).

The plant community growing on a particular soil is regarded as an important factor to explain the distribution of dorylaims (Weischer, 1975; Liébanas *et al.*, 2004). Thus, the nature (perennial/deciduous, close/open) and age of forests affect their number (Bassus, 1962; Niblack & Bernard, 1985; McSorley, 2013), and non-forest habitats present higher populations of *Eudorylaimus* and *Mesodorylaimus* (McSorley, 2013). Other authors provided more detailed results. For example, Háněl (2008) observed that *Oxydirus oxycephalus*, *Tylencholaimellus striatus* and *Tylencholaimus constrictus* were more abundant in deciduous forests (*Alnus glutinosa*, *A. incana*, *Quercus robur* and *Tilia cordata*) than in coniferous forests in restored soils on colliery spoils in the Czech Republic. Cesarz *et al.* (2013) noted that several dorylaimid taxa were detrimentally affected by lime (*Tilia* spp.), but the negative effect was reduced by beech (*Fagus sylvatica*) in a deciduous forest in Central Germany. Crop type also conditions the species distribution as observed by Murphy *et al.* (1974) in *Xiphinema americanum* populations of North America.

Vertical (depth) distribution

Some dorylaims show relevant differences in the number of individuals when their vertical (depth) distribution is analysed. Thorne (1930) was probably the first author who provided interesting information about this matter when analysing the distribution of four nygolaims (Nygolaimina) species in American soils. This author found that there is a general pattern of decreasing abundance with increasing depth, but also that one of the four species, *Nygolaimus tenuis* in his study, differed from the remaining three in its vertical distribution as it occurred only at medium to low depths (Fig. 13.1). The decreasing pattern of abundance in vertical distribution of dorylaims was later reported in both free-living (Castillo *et al.*, 1985) and plant-parasitic (Flegg, 1968; Cohn, 1969; Boag *et al.*, 1987; Weischer & Almeida, 1995) forms (Fig. 13.2A). Nevertheless, other patterns also occur. An abundance peak at medium depths was observed by Flegg (1968) in *Xiphinema* sp. at 50 cm, and in *Longidorus macrosoma* at 70 cm (Fig. 13.2B). More exceptional is the regular increasing abundance with depths to about 75 cm and sometimes beyond, as reported in *Longidorus profundorum* by Flegg (1968; Fig. 13.2C) and in *Longidorus africanus* by Ploeg (1998). Nagy *et al.* (1998), however, did not find any definite vertical distribution pattern for *Xiphinema vuittenezi*. Some of these patterns have been confirmed by means of statistical methods. Thus, Andrés & Arias (1982) found a significantly positive correlation of *L. profundorum* with the deepest (25-50 cm) level of vertical distribution analysed, significantly negative correlation with the most surface level (0-10 cm), and no correlation with the intermediate level (10-25 cm).

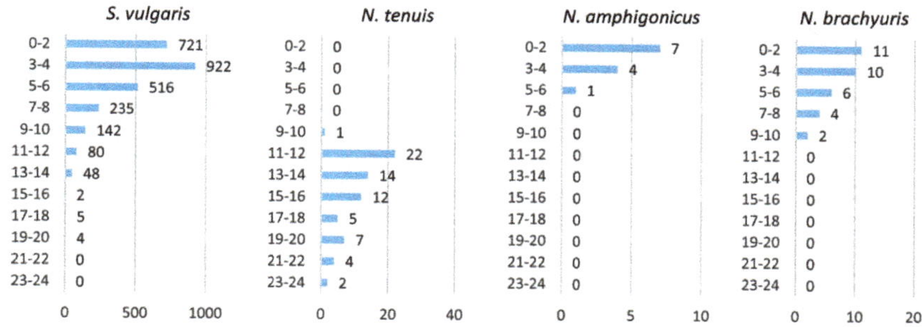

Fig. 13.1. *Vertical distribution of four nygolaimid species in North American soils:* Solididens vulgaris, Nygolaimus tenuis, N. amphigonicus *and* N. brachyuris. *(Depth (vertical axis) in inches, abundance (horizontal axis) in number of individuals; from Thorne's (1930) original data.)*

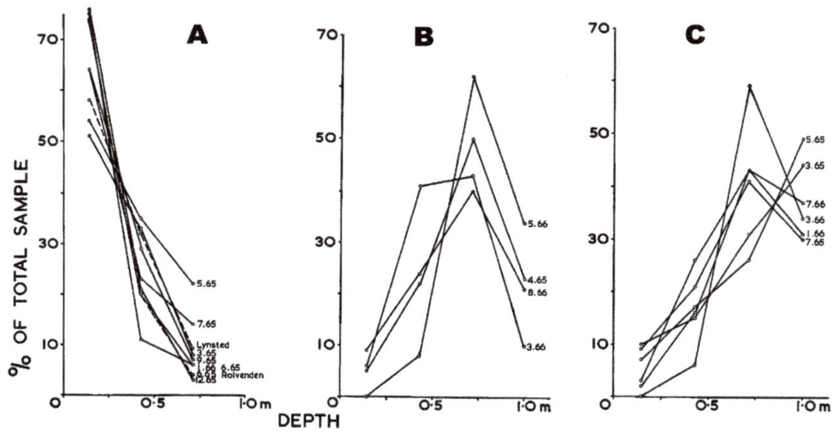

Fig. 13.2. *Vertical (depth) distribution of three species of Longidoridae. A:* Xiphinema diversicaudatum*; B:* Longidorus macrosoma*; C:* L. profundorum. *(After Flegg (1968),* Nematologica *14, pp. 192-193, Figs 2, 3.0.)*

Vertical distribution of dorylaims can be influenced by other environmental variables. Feil *et al.* (1997) observed that soil properties affect the distribution of *Xiphinema index*. MacGuidwin (1989) detected seasonal changes in the distribution of *Longidorus breviannulatus*, with higher abundance in depths of 0-15 cm during spring season and in depths of 15-30 cm during fall season, a redistribution explained as due to migration. Conversely, Krnjaic (1971), who studied the vertical distribution of

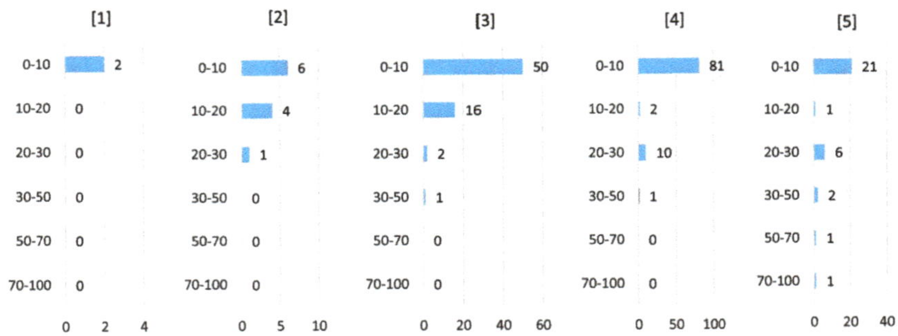

Fig. 13.3. *Vertical distribution of* Longidorella deliblatica *in five habitats (1 = bar sand dune, 2 =* Coryspermeto-Polygonetum arenariae, *3 =* Festucetum vaginatae fumanetosum, *4 =* F. vaginatae deliblaticum muscetosum, *5 =* Crysopogonetum pannonicum typicum*) from the former Yugoslavia. (Depth (vertical axis) in cm, abundance (horizontal axis) in number of individuals; from Krnjaic's (1971) original data.)*

Longidorella deliblatica in five habitats of the former Yugoslavia (Fig. 13.3), found a similar pattern in all the cases, with only small differences between them.

Temporal (seasonal) dynamics

Available data suggest that the seasonal distribution of dorylaimid populations may follow some recognisable patterns. A bimodal scheme has been reported in both free-living (Gantait *et al.*, 2010, 2011a, b) and plant-parasitic (Griffin & Darling, 1964; Krnjaic, 1971) forms, with a first peak occurring in spring and a second one in autumn (Fig. 13.4), often following the development of plants (Krnjaic, 1971). Feil *et al.* (1997), however, found that *Xiphinema index* has two maximum values (winter and spring) in Californian vineyards. A unimodal pattern has been observed in *Carcharolaimus banaticus* found in sand dune habitats in the former Yugoslavia (Krnjaic & Loof, 1975; Fig. 13.5) and *Paractinolaimus microdentatus* dwelling in soils of an ash forest (Castillo *et al.*, 1988), with maximum abundance observed in the spring and summer periods, respectively. Several authors (Andrés & Arias, 1989; Philis, 1996; Ploeg, 1998) found a similar scheme for *Longidorus* and *Xiphinema* species, with maximum population levels in spring-summer and minimum ones in winter, thus being largely correlated with soil temperature. More complex patterns may also occur (Prota & Garau,

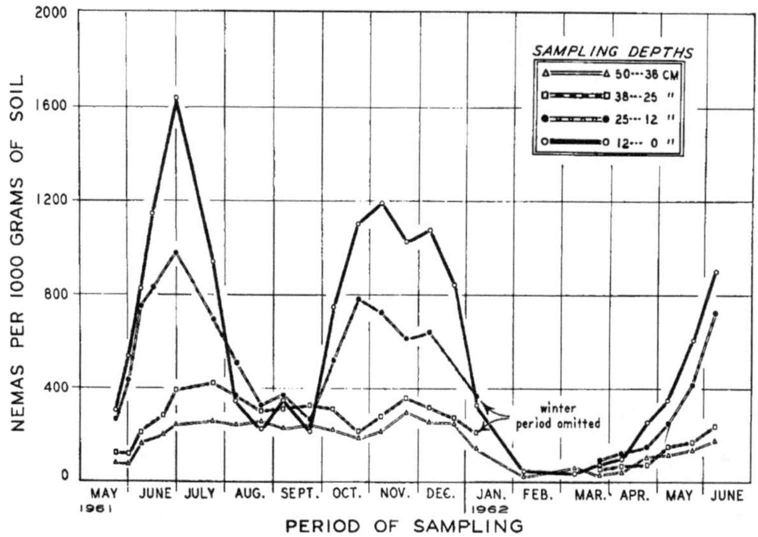

Fig. 13.4. *Temporal (seasonal) variation of* Xiphinema americanum *population (after Griffin & Darling (1964),* Nematologica *10, p. 473, Fig. 1).*

1973). Nath *et al.* (1998) observed three peaks (March, July and October) in the annual population variation of *Xiphinema radicicola* in pineapple plantations in India.

Regional distribution

Attempts at analysing the distribution of dorylaims at a medium (regional) scale, understood as being extensive areas often including many kinds of habitats, are few and the results obtained do not allow either the detection of common general patterns or the elucidation of the processes driving them. Nonetheless, some contributions deserve a mention.

The distribution of Longidoridae species in Europe was studied by Dalmasso (1970), who classified them into four ecological (geographical) groups: *i*) Atlantic; *ii*) Continental; *iii*) Mediterranean; and *iv*) probably recently introduced. Navas *et al.* (1990) studied the distribution of the same dorylaimid group in the Euromediterranean area, and identified some remarkable chorological relationships, either chorotypes (species with similar distribution) or biotic regions (areas with species in com-

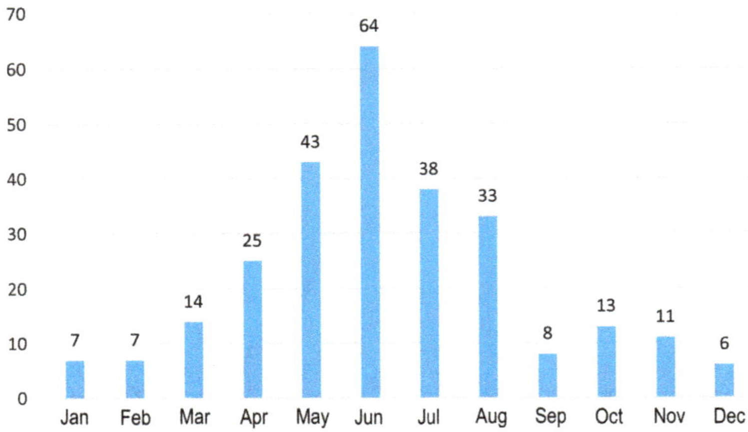

Fig. 13.5. *Temporal (seasonal) variation of* Carcharolaimus banaticus *in sand dune of the former Yugoslavia. (Abundance (vertical axis) in number of individuals, month (horizontal axis); after Krnjaic & Loof's (1975) original data.)*

mon; Fig. 13.6). Liébanas *et al.* (2002) analysed the distribution of 138 species of dorylaims and mononchs in 203 soil samples collected in a protected area (natural park) of the south-eastern Iberian Peninsula and, based on their absence/presence data, detected 14 collective chorotypes and a dozen individual species with an identifiable and characterisable distribution. Velasco-Castrillón *et al.* (2014) provided a review of the species hitherto recorded in Antarctica, which consists of 19 dorylaimid species, 11 of them belonging to the genus *Eudorylaimus* and four to *Mesodorylaimus*.

The processes involved in the origin of these distribution patterns remain obscure in many cases, but valid tentative hypotheses have been proposed to explain them. Most of these hypotheses consider not one, but a set of ecological and/or historical causes that determine the observed distributions. Dalmasso (1970) found that a combination of environmental factors (climate, plant and dispersal) defined his four ecological groups of European longidorid species (see above). Vegetation (plant community) and soil properties influence the distribution of longidorid species not only at the local, but also at the regional scale in Europe (Weischer & Almeida, 1995). For example, the nature of soils, derived from either calcareous or siliceous rocks, seems to be able to explain the ranges of several forms. Navas *et al.* (1993) studied the distribution of

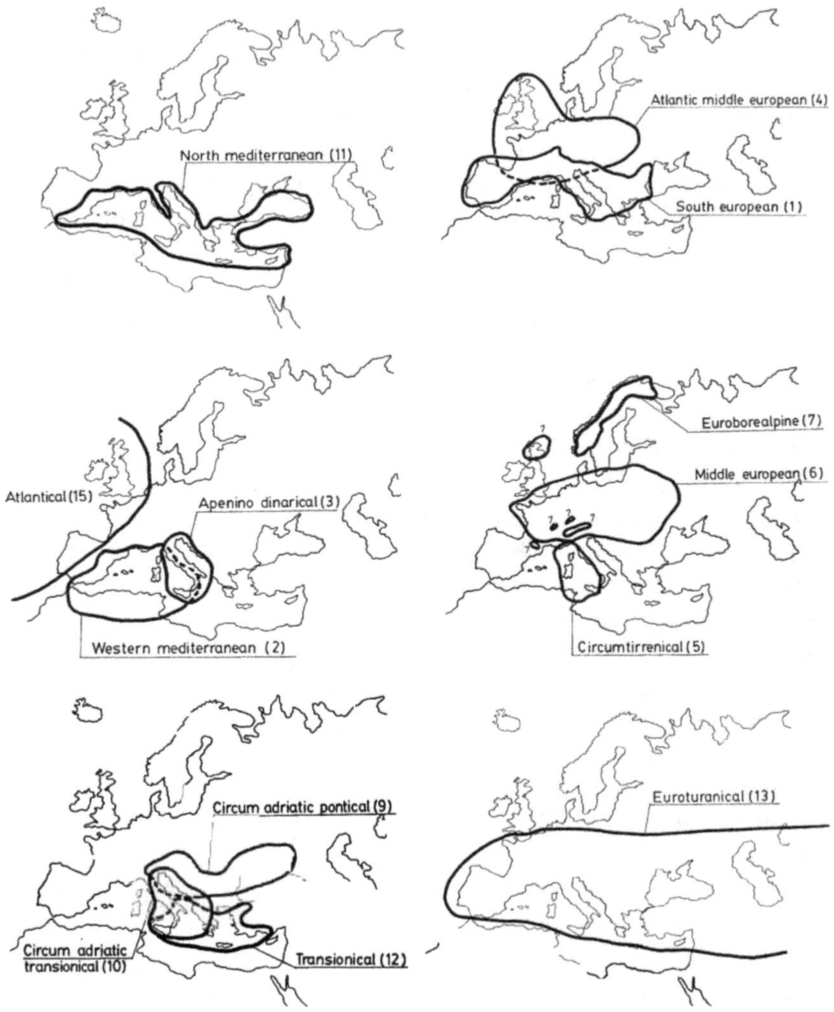

Fig. 13.6. *Mediterranean biotic regions based on the distribution of longidorid species. (See text and the original contribution by Navas* et al. *for additional explanation.) (After Navas* et al. *(1990),* Nematologia Mediterranea *18, p. 105, Fig. 7.)*

Longidorus species in the Euromediterranean area and concluded that vicariant biogeography can resolve some aspects of the current pattern of their distribution and speciation origin, but also that dispersal is a primary explanation for many of them. Liébanas *et al.* (2004) concluded that up to seven (but very often 1-3) variables may explain the spatial distribution of their collective and individual chorotypes, but also that a particular variable may largely differ in its importance in explaining the observed distribution patterns.

The effect of quaternary glaciations explains some regional patterns such as the present distribution of longidorids in France (Dalmasso, 1970), the scarcity of longidorids in Northern Europe (Topham & Alphey, 1985) or the distribution of longidorids in the Euromediterranean region (Navas *et al.*, 1993). The highly diverse faunas of longidorid forms in the eastern Mediterranean territories were attributed to Miocene plate tectonics in that area (Topham & Alphey, 1985).

Global distribution

The study and analysis of worldwide (global) distribution of dorylaims is a true scientific challenge, with serious handicaps to be overcome, in particular, the absence of robust information concerning the faunas of many regions. As stated by Coomans *et al.* (2001), regarding the geographical distribution of the speciose genus *Xiphinema*, there are practical problems when mapping the distribution of the species: *i*) lack of information from countries for which nematological surveys have not been undertaken yet; *ii*) insufficient information in the literature to enable unambiguous confirmation or identifications; *iii*) large variation in the number and extent of surveys among the surveyed countries; and *iv*) passive dispersal of species due to transport of rooted plants and soil.

These problems become especially serious in the case of free-living dorylaims – faunistic studies consume much effort and resources and need a solid taxonomical background that is often underestimated if not entirely removed from the scientific priorities of many research programmes. Taking into account the fact that dorylaims are microorganisms or members of the meiofauna, it is necessary to confront some simple ideas or aphorisms, not always scientifically tested; for example, the statement "Everything is everywhere, but the environment selects", attributed to both Beijerinck and his pupil Becking (de Wit & Bouvier, 2006; Zullini, 2014) has been repeatedly cited. According to Zullini (2018), the biogeography of nematodes reflects their intermediate position between macro- and microorganisms. Very exceptionally, however, a few contributions have provided valuable information about world fau-

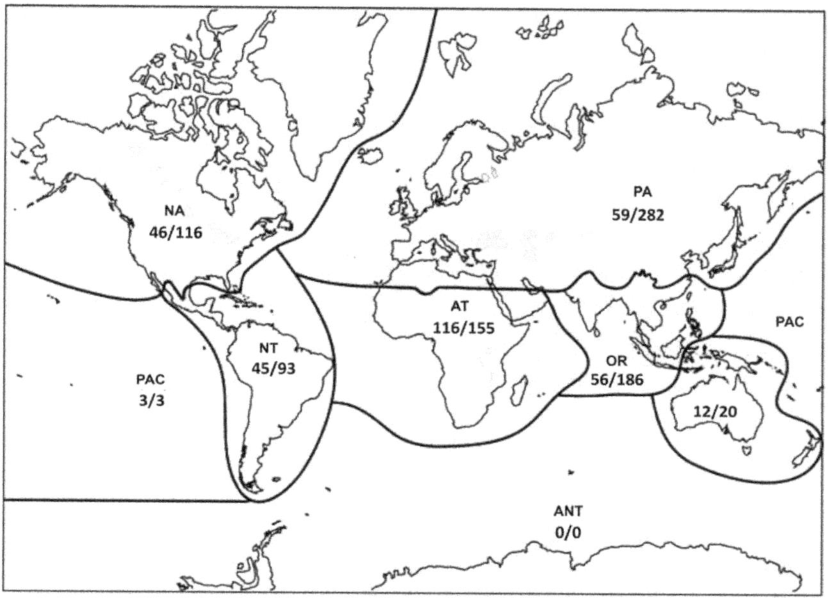

Fig. 13.7. *Distribution of freshwater dorylaims (genera number/species number) per zoogeographical region (adapted from Abebe* et al. *(2008)).*

nas, often associated with particular habitats. Coomans *et al.* (2001) analysed the distribution of *Xiphinema* (234 valid) species and suggested some general but interesting statements or hypotheses: *i*) many species are concentrated in tropical and African territories; *ii*) most species are known to occur only in their type locality; *iii*) the wider distribution of other species could be due to several causes; and *iv*) disjunct distribution patterns are often difficult to explain.

Decraemer & Coomans (1994) compiled the nematode fauna recorded in ancient lakes (among others, Baikal, Tanganyika, Ohrid, Kinneret), where 13 dorylaimid (out of 53 nematode) genera and 19 (out of 85) species were present, showing a high level of endemicity, as 1-3 species were endemic in each lake. Abebe *et al.* (2008) analysed the distribution of freshwater nematodes in the zoogeographical regions (Fig. 13.7), found that 103 dorylaimid genera (out of 319 nematode genera) and 610 (out of 1808) species constituted the world fauna, and provided interesting ideas about their biogeography. Du Preez *et al.* (2017) compiled the nematode fauna occurring in caves around the world, listed 22 dorylaimid genera (out of 109 nematode genera) and

33 dorylaimid species (out of 171 nematode species), and proposed a classification of cave-dwelling forms.

THE DISTRIBUTION OF PARTICULAR TAXA – SOME EXAMPLES

It is totally justified to assume that some dorylaimid species may be ubiquitous or cosmopolitan. Thus, *Aporcelaimellus obtusicaudatus* is probably the most frequent and abundant dorylaimid species in natural soils everywhere (Andrássy, 2009). Coomans *et al.* (2001) have summarised the reasons for this widespread nature of some forms: *i*) species antiquity; *ii*) more generalistic life strategy; *iii*) higher (passive) dispersive capacity; and *iv*) paleogeography of the ancestral centre of origin.

Conversely, many other dorylaimid groups are known to occur in more restricted ranges. Some examples may give an idea of particular distributions (Fig. 13.8). The eight known species of the genus *Pachydorylaimus* inhabit Neotropical soils of Central America and the northern part of South America. The seven representatives of *Boreolaimus* only occur in Arctic territories. The members of the genera *Enchodelus* (24), *Heterodorus* (25) and *Rhyssocolpus* (eight) are typical components of septentrional (Northern Hemisphere) faunas. The 35 *Oriverutus* species dwell in tropical soils worldwide, with just a few records in northern temperate countries (Peña-Santiago *et al.*, 2014). All members of the genera *Australodorus*, *Paraxiphidorus* and *Xiphidorus* (13 species in total), are restricted to riparian habitats in three South American countries of the South Atlantic (Argentina, Brazil and Uruguay).

HISTORICAL BIOGEOGRAPHY

A few contributions have analysed the influence of historical processes on the current spread of some dorylaimid species. In a series of contributions, V.R. Ferris and her collaborators (1976, 1980, 1981, 1983) discussed the cases of several belondirid and leptonchid taxa whose geographical distribution may be tentatively explained by plate tectonics (also known as continental drift). Ferris (1979; see also Ferris *et al.*, 1981) hypothesised that leptonchid forms derived from a Pangaean ancestor with subsequent radiation primarily in areas that were formerly part of the Gondwanian supercontinent. Ferris *et al.* (1980) suggested very interesting ideas about the phylogeny and biogeographical distribution of the genera *Tarjanius* and *Oxydirus*, with ancestors in the Afro-

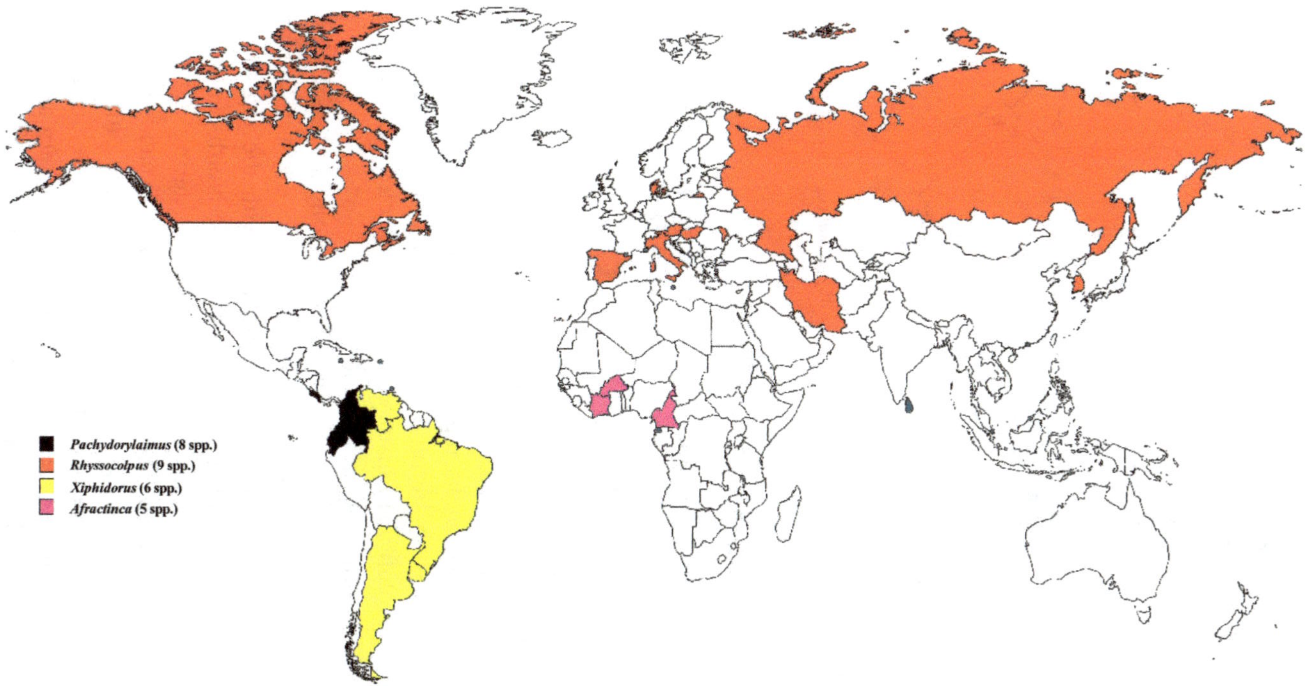

Fig. 13.8. *World geographical distribution of four selected dorylaimid genera showing restricted spread:* Afractinca *(pink colour),* Pachydorylaimus *(green),* Rhyssocolpus *(red) and* Xiphidorus *(yellow).*

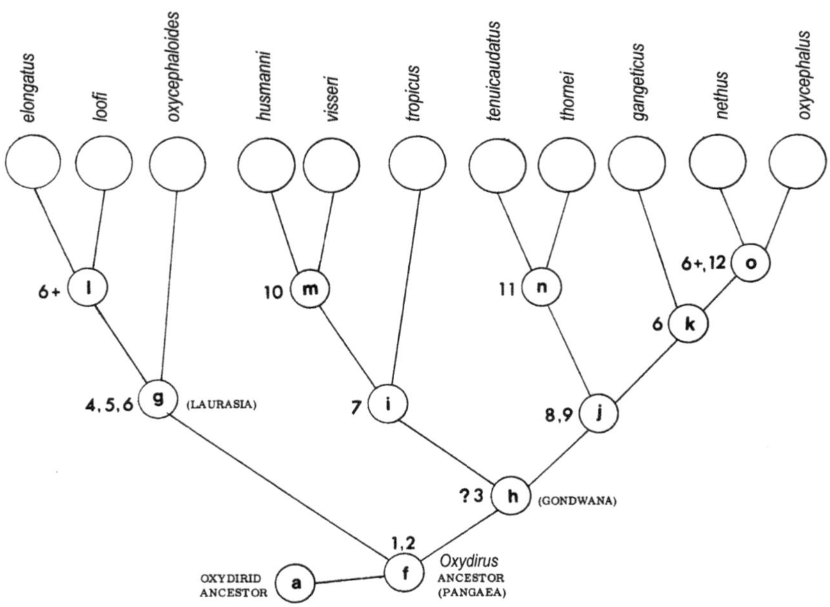

Fig. 13.9. *Cladogram of* Oxydirus *species, with common hypothetical Pangaean, Laurasian and Gondwanian ancestors. (After Ferris* et al. *(1980),* Purdue University Agricultural Experiment Station Research Bulletin *965, p. 27, Fig. 19.)*

Indian part of Gondwanaland and Pangaea, respectively (Fig. 13.9). The origin of four *Belondira* species characterised by having an especially swollen cuticle in the caudal region can be hypothetically traced from a Gondwanian ancestor (Ferris *et al.*, 1983; Fig. 13.10). Ferris (1983) also traced the evolutionary history of *Tyleptus* species according to the tectonic divisions affecting the continents.

Coomans (1985) explained the evolution of Longidoridae following the drift of tectonic plates and paleogeographic events. The origin of this group probably took place, correlated with the *Glossopteris*-flora, in northern Gondwanaland (present West-, North- and Central Africa and the northern half of South America) during the Permian period (280-225 MYBP). The genus *Xiphinema* expanded its range to the southern parts of Gondwanaland. During the Jurassic, other longidorids may have appeared in association with more diverse floras, including angiosperms, *i.e.*, *Paralongidorus*, in the eastern and northern parts of Gondwanaland, and then much later invading Europe. The

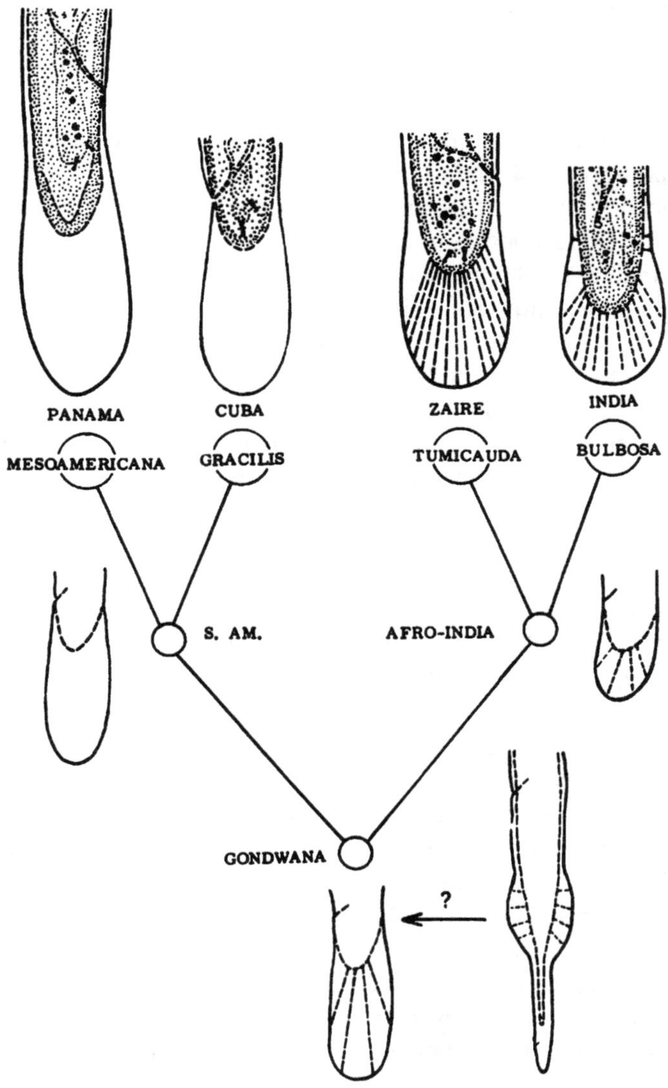

Fig. 13.10. *Phylogenetic diagram of relationships hypothesised for four species of* Belondira *with very long tail and very long cuticular expansion. (After Ferris* et al. *(1983),* Purdue University Agricultural Experiment Station Research Bulletin *979, p. 46, Fig. 22.)*

radiation of *Longidoroides* was limited and its later dispersal confined to southern and southwest Africa, with the possibility for some dispersal to South America through the then existing connection between southwest Africa and South America. *Longidorus* also originated before the break-up of Gondwanaland, and dispersed in all directions except towards South America, radiating most in Laurasia, especially in Europe. *Paraxiphidorus* and *Xiphidorus* are confined to South America as they probably originated in this continent after the break-up of Gondwanaland during the Cretaceous period (100 MYBP).

Vinciguerra (1988) pointed out that actinolaims (Actinolaimidae) also probably have a Jurassic origin in Gondwanaland, with an early radiation before the breakup of this paleocontinent as some genera are very widespread, mostly in the tropical regions of Africa, America and Asia. As the greatest number of actinolaimid genera and species, including the most plesiomorphic forms, are found in Africa, the same author suggested that the early radiation took place in this continent.

Phylogeography

A modern approach to historical biogeography, based on the analysis of gene genealogy, and termed phylogeography (Avise, 2000), has gained major interest over the last decades. Once again, the lack of information concerning the geographical distribution of nematodes, and especially of dorylaims, is a serious handicap to obtaining relevant results on this matter. Nevertheless, there are some interesting contributions that show recognisable general patterns and processes. Thus, Chizhov *et al.* (2014) analysed the distribution of European populations of *Xiphinema diversicaudatum* and hypothesised that during the last glacial maximum the species probably persisted in refuge areas in the Carpathian mountains or the Alps, later expanding and colonising other regions in Europe. More recently, Nguyen *et al.* (2019) analysed the world distribution of *Xiphinema index* using several molecular markers and distinguished two clades, each with two subclades, all having a marked geographical distribution (Fig. 13.11). Thus, populations from the eastern Mediterranean and the Near and Middle East were grouped into three of these subclades, whereas samples from the western Mediterranean, Europe and the Americas all belonged to the fourth subclade. These authors also provided interesting evidence about the native area of this species and its later dispersion during Antiquity.

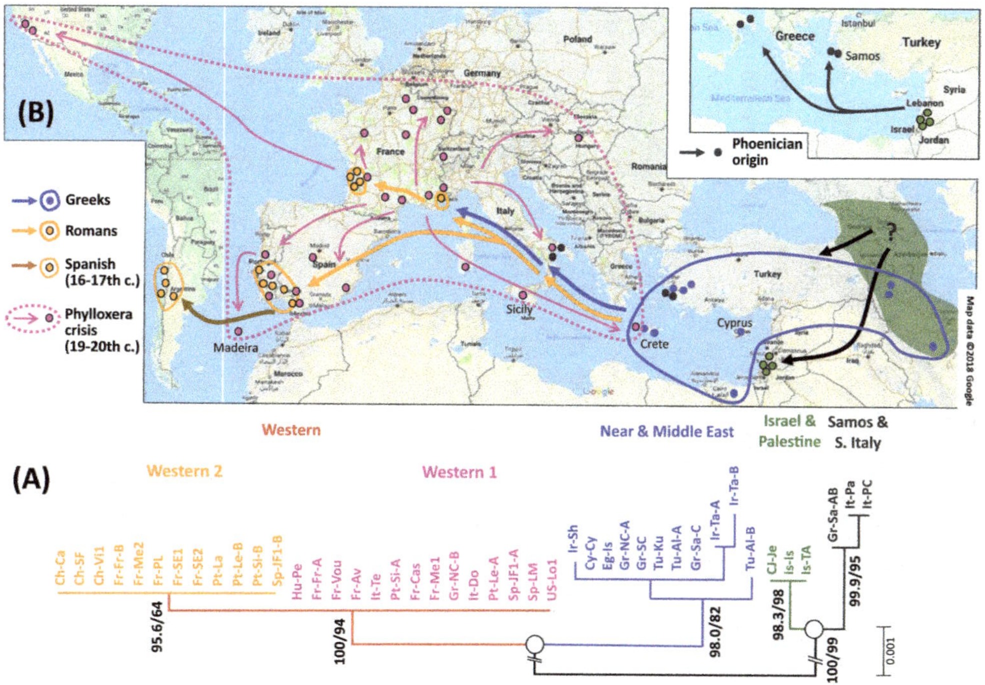

Fig. 13.11. *Phylogeography of* Xiphinema index *based on the analysis of several molecular markers. Populations from the eastern Mediterranean and the Near and Middle East are grouped into three subclades, whereas the samples from the western Mediterranean, Europe and the Americas form the fourth subclade. (After Nguyen* et al. *(2019),* Scientific Reports *9, 7313, p. 5, Fig. 2.)*

Dispersal and transport

Like many other nematodes, dorylaims are able to move in soil by mean of active locomotion (Pitcher, 1975), depending on the physical and chemical properties of their habitat. Harrison & Winslow (1961) estimated that *Xiphinema diversicaudatum* spread from a hedge into woodland at an average rate of 30 cm per year over a period of 75 years. And Thomas (1981) noted that *Longidorus elongatus* and *X. diversicaudatum* moved through a sandy loam soil in both vertical and horizontal planes at rates of 7 and 10 cm per month, respectively.

Passive transport of free-living nematodes, including dorylaims, by abiotic and biotic agents has been documented by a few authors, who assumed that they represented key pathways for dispersal and colonisation of new areas. Waliullah (1984, 1989) and Roccuzzo & Ciancio (1991) observed this process as driven by irrigation waters, and Villenave *et al.* (2003) by run-off water. On the other hand, Baujard & Martiny (1994) reported the same process driven by the wind. Phoretic relationships between the dorylaimid species *Nygolaimoides borborophilus* and five species of flies of the families Sphaeroceridae and Sepsidae, collected over cow dung in pastures, were noted by Chizov (1996), the flies presumably acting as a transport mechanism to disperse the nematodes to other areas.

Special mention must be made of the dispersal of dorylaimid species due to anthropogenic actions, very often associated with exchange or trade of cultivated, forest and ornamental plants. Thus, Ferris (1983) suggested that the presence of vicariant and similar populations/species of the genus *Dorylaimellus* in very distant areas was (p. 156) "...a highly probable case of relatively recent dispersal from Europe to New Zealand", whereas Boag & Brown (1987) hypothesised that *Longidorus vineacola*, a widespread species in Western Europe, but very occasionally found in Ireland and Scotland, was introduced into the British Isles by humans.

Chorologial relationships

The geographical distributions of species can be compared and analysed to establish their relationships. Particularly interesting is the detection of chorotypes, *i.e.*, two or more species that share a similar distribution pattern significantly different from the patterns observed in other species (Baroni-Urbani *et al.*, 1978). Zullini (1970) was probably the first author to mention the existence of this kind of relationship in some dorylaimid species, either among them and/or with other non-dorylaimid nematodes. Later, several authors observed similar cases in members of the family Longidoridae. Bello *et al.* (1986) classified the Iberian species in eight groups according to their geographical distribution. De Waele & Coomans (1990) found differences in the association of longidorid nematodes in Britain and Belgium, probably a consequence of adaptations to changing habitats by geographically separate populations of the same species. Navas *et al.* (1990) clustered Mediterranean species into eight faunistic elements (= chorotypes). Zoon *et al.* (1993) found that *Longidorus dunensis* and *Pratylenchus* spp. presented a correlated distribution in sand dunes. Marshall *et al.* (1998) noted that *Longidorus elongatus* and *L. goodeyi* displayed a correlated distribution at scales up to about 20 cm. Liébanas *et al.* (2002) identified several dorylaimid chorotypes that were environmentally characterised.

Conversely, two or more species of the same genus may spread in separate (allopatric) or dissimilar (antipatric or excluding) geographical ranges. In the first case, several species that share a recent common ancestor occupy more or less distant areas as a result of vicariant processes. In the second case, two species display opposite ecological behaviours or requirements (climatic conditions, habitat type, soil features, *etc.*). For example, several *Xiphinema* species display nearly exclusive ranges in the Iberian Peninsula (Navas & Arias, 1986), and Navas *et al.* (1988) observed that *X. diversicaudatum* and *X. pachtaicum*, the former a typical inhabitant of the Eurosiberian region, the latter a species nearly restricted to Mediterranean territories, occupied dissimilar geographical ranges.

Biotic interactions

Many aspects of the relationships among dorylaims and other living beings remain poorly explored or simply unknown. Nevertheless, some achievements in this field deserve attention.

Several representatives of the plant-feeding dorylaimid genera *Longidorus, Paralongidorus* and *Xiphinema* are able to transmit nepoviruses, small isometric particles, with a variable vector specificity (Brown *et al.*,

1988, 1993, 1998; Taylor & Brown, 1997). The transmission process involves four steps (Harrison *et al.*, 1974; Taylor & Robertson, 1975): *i*) uptake of the virus particles from an infected plant; *ii*) survival of infective particles in the nematode; *iii*) inoculation by the nematode to another plant; and *iv*) infection of the receptor plant by the virus.

The presence of bacterial endosymbionts in the epithelia of the ovary wall of mature females of longidorid species has been repeatedly reported (see revision by Palomares-Rius *et al.*, 2016). These prokaryotic organisms are transmitted vertically (Coomans *et al.*, 2000) following a relatively complex process: *i*) in the intra-uterine egg, the bacteria migrate to one of the poles and eventually surround the host nucleus of the intestinal stem cell during the phase that corresponds to the embryonated egg; *ii*) in juveniles, they occur only in the gut epithelial cells; and *iii*) they invade the growing oocytes before the last moult (Vandeckerckhove *et al.*, 2002). The bacteria involved belong to the genus '*Candidatus* Xiphinematobacter' and the family Burkholderiaceae, and have several species of the genus *Xiphinema* as their hosts. Unfortunately, the biology of this interaction has not yet been elucidated, but a high degree of phylogenetic congruence of the bacteria with their host has been revealed. Coomans & Willems (1998) suggested that the presence of bacteria in species of the *X. americanum*-group may induce sex ratio distortion and parthenogenesis, a characteristic feature of these species. The association between endospore-forming and mycelial bacteria of the genus *Pasteuria*, in particular, the *P. penetrans* group, and nematodes has received much attention, mainly due to the use of the former as potential agents for biological control of the latter. *Pasteuria* sporangia can be found attached only to the cuticle and/or in the pseudocoel of many nematode taxa, including free-living and plant-parasitic dorylaims (Sayre & Starr, 1988; Ciancio, 1995). Thus, Sturhan (1988) listed 17 species of dorylaims, including both plant-parasitic and free-living forms, parasitised by this kind of bacterium in samples collected from Europe (Bulgaria, Germany, UK-England), the Azores, and Nicaragua.

Interactions between dorylaims and protozoan forms seem to be much more occasional, but they do occur. Morone de Lucia & Grimaldi de Zio (1973) observed the presence of microsporidia in *Xiphinema diversicaudatum* gonads, and Tahseen *et al.* (2003) reported and illustrated the association of *Vorticella* specimens with an *Aporcelaimus* species, although the precise nature of the relationship was not established.

There is experimental evidence for certain nematophagous fungi attacking dorylaims. Nevertheless, this kind of association has not received much attention. Boag *et al.* (1988) observed that, out of several species of longidorid dorylaims, only *Longidorus attenuatus* suffered attack by *Arthrobotrys dasguptae*, with adhesive knobs confined to the anterior and caudal regions. Jaffee (1986), however, observed that four zoosporic fungi were able to infect *Xiphinema americanum* and *X. rivesi*. Ciancio *et al.* (1986) observed *Hirsutella rhossiliensis* parasitising *Xiphinema diversicaudatum*, and Ciancio & Chinappen (1987) found that *Paralongidorus buchae* is a host for *Catenaria anguillulae*.

Many arthropods, mainly mites, are able to feed on nematodes, including dorylaims (Walter & Ikonen, 1989; Bilgrami & Tahseen, 1992; Bilgrami, 1994, 1997).

Dorylaims as bioindicators in a global change scenario

The dorylaims are considered to be persisters and/or K-strategists and, in general, omnivorous or predatory. They have an especially permeable cuticle and are thus very sensitive to pollutants and other environmental perturbations and stresses (Tenuta & Ferris, 2004). All of these biological features make the dorylaims excellent bioindicators for monitoring purposes, especially for assessing the environmental quality (health) of the habitats in which they dwell (Wasilewska, 1997; Murialdo *et al.*, 2002; Todd *et al.*, 2006). Thus, the response of dorylaims to pollution by heavy metals results in a significant decrease in their number or their total disappearance, as demonstrated by Nagy (1999; see also Ekschmitt & Korthals, 2006), who found that dorylaimid forms were totally absent in experimental plots treated with Se and often absent in plots treated with Cr, underlining the negative effects of these pollutants. Dorylaims represent the most sensitive nematode trophic group (omnivores) when different management or manipulation strategies are implemented in both natural (Ruess *et al.*, 2002) and cultivated (Wang *et al.*, 2006; McSorley, 2012) soils. Similarly, Neher & Weicht (2013) reviewed and analysed the responses of the soil nematode community to increasing CO_2 levels and found significant differences in the behaviour of several genera, including dorylaims, with *Tylencholaimus* abundance significantly decreasing in pine forests. Conversely, but also confirming their bioindicator usefulness, some dorylaimid genera significantly increase their abundance in restored ecosystems (Todd *et al.*, 2006; Briar *et al.*, 2012).

The effect of climatic changes on the spatial distribution of dorylaims has received less attention. Nearly 30 years ago, however, Boag *et al.* (1991) found a close association between temperature and geographical distribution of several European longidorid species, and hypothesised that global warming might result in an increase of the populations of these species in northern territories and the spread of nematodes from the south at a slow rate, probably depending on dispersal processes induced or facilitated by human activities.

Notes on survival

The ability of dorylaims to survive for long periods under unfavourable or extreme conditions has hardly been explored in free-living forms, but relatively well documented in longidorid species. McNamara (1980; see also Stirling, 1985) compiled the available information concerning *Xiphinema* and *Longidorus* species and reported that they can survive for a few months to several years (maximum 54 months) in plant-free field soil. The same author suggested that some factor (probably a microorganism) induced partially quiescence (and survival) in *X. diversicaudatum*.

Anhydrobiosis is probably a survival strategy in *Eudorylaimus* species against adverse environmental conditions (Lee, 1961; Wall & Virginia, 1999). As happens in other nematodes, dorylaims may form a coiled spiral, significantly reducing the body surface exposure. Once the critical period is over, the nematodes need only a few hours to recover, becoming active 24 h after their re-hydration (Freckman *et al.*, 1977).

Under experimental conditions, survival of *X. diversicaudatum* in different pH gradients, mineral salts and fatty acids shows significant variations (Malik & Jairajpuri, 1984): the optimum pH is 6.0-6.6, adults survived longer than juveniles in both salts and acids, and the first juvenile stage survived for the shortest time. Sultan & Ferris (1991) found that moisture and soil particle size had a marked effect on the survival and population increase of *X. index*.

References

ABEBE, E., DECRAEMER, W. & DE LEY, P. (2008). Global diversity of nematodes (Nematoda) in freshwater. *Hydrobiologia* 595(2007), 67-78. DOI: 10.1007/s10750-007-9005-5

ANDRÁSSY, I. (2009). *Free-living nematodes of Hungary, III (Nematoda errantia)*. In: *Pedozoologica Hungarica 5* (Series Editors: Csuzdi, C. & Mahunka, S.). Budapest, Hungary, Hungarian Natural History Museum and Systematic Zoology Research Group of the Hungarian Academy of Sciences.

ANDRÉS, M.F. & ARIAS, M. (1982). Distribución vertical de los nematodos del género *Longidorus* en la región central. *Boletín del Servicio de Plagas* 8, 143-148.

ANDRÉS, M.F. & ARIAS, M. (1989). Observations on the population dynamics of *Longidorus belloi* in cereal fields of the Central Region (Spain). *Nematologia Mediterranea* 17, 35-37.

AVISE, J. (2000). *Phylogeography: the history and formation of species.* Cambridge, MA, USA, Harvard University Press.

BARONI-URBANI, C., RUFFO, S. & TAGLIANTI, A.F. (1978). Materiali per una biogeografía italiana fondata su alcuni generi di coleotteri cicindelidi, carabidi e crisomelidi. *Estratto dalle Memorie della Societa Entomologica Italiana* 56, 35-92.

BASSUS, W. (1962). Über die Vertikalverteilung und den Massenwechsel der Nematoden in Waldböden Mitteldeutschlands. *Nematologica* 7, 281-293. DOI: 10.1163/187529262X00576

BAUJARD, P. & MARTINY, B. (1994). Transport of nematodes by wind in the peanut cropping area of Senegal, West Africa. *Fundamental and Applied Nematology* 17, 543-550.

BELLO, A., TOPHAM, P.B., ALPHEY, T.J.W. & DALE, A. (1986). Biogeographical classification of some plant-parasitic nematode species groups in Spain. *Nematologia Mediterranea* 14, 55-72.

BILGRAMI, A.L. (1994). Predatory behaviour of a nematode feeding mite *Tyrophagus putrescentiae* (Sarcoptiformes: Acaridae). *Fundamental and Applied Nematology* 17, 293-297.

BILGRAMI, A.L. (1997). Evaluation of the predation abilities of the mite *Hypoaspis calcuttaensis*, predaceous on plant and soil nematodes. *Fundamental and Applied Nematology* 20, 96-98

BILGRAMI, A.L. & TAHSEEN, Q. (1992). A nematode feeding mite, *Tyrophagus putrescentiae* (Sarcoptiformes: Acaridae). *Fundamental and Applied Nematology* 15, 477-478.

BOAG, B. & BROWN, D.J.F. (1987). The occurrence of *Longidorus vineacola* in Scotland with notes on its distribution in Europe. *Nematologia Mediterranea* 15, 51-57.

Boag, B., Brown, D.J.F. & Topham, P.B. (1987). Vertical and horizontal distribution of virus-vector nematodes and implications for sampling procedures. *Nematologica* 33, 83-96. DOI: 10.1163/187529287X00236

Boag, B., Robertson, W.M. & Ainsworth, L.F. (1988). Observations on the specificity of the nematophagous fungus *Arthrobotrys dasguptae* (Shome & Shome) to plant-parasitic nematodes. *Nematologica* 34, 238-245. DOI: 10.1163/002825988X00332

Boag, B., Crawford, J.W. & Neilson, R. (1991). The effect of potential climatic changes on the geographical distribution of the plant-parasitic nematodes *Xiphinema* and *Longidorus* in Europe. *Nematologica* 37, 312-323. DOI: 10.1163/187529291X00312

Bongers, T. (1990). The maturity index: an ecological measure of environmental disturbance based on nematode species composition. *Oecologia* 83, 14-19. DOI: 10.1007/BF00324627

Bongers, T. & Ferris, H. (1999). Nematode community structure as a bioindicator in environmental monitoring. *Trends in Ecology & Evolution* 14, 224-228. DOI: 10.1016/S0169-5347(98)01583-3

Bongers, T., de Goede, R.G.M., Korthals, G. & Yeates, G.W. (1995). Proposed changes of c-p classification for nematodes. *Russian Journal of Nematology* 3, 61-62.

Briar, S.S., Barker, C., Tenuta, M. & Entz, M.H. (2012). Soil nematode responses to crop management and conversion to native grasses. *Journal of Nematology* 44, 245-254.

Brown, D.J.F. & Weischer, B. (1998). Specificity, exclusivity and complementarity in the transmission of plant viruses by plant parasitic nematodes: an annotated terminology. *Fundamental and Applied Nematology* 21, 1-11.

Brown, D.J.F., Lamberti, F., Taylor, C.E. & Trudgill, D.L. (1988). Nematode-virus plant interactions. *Nematologia Mediterranea* 16, 153-158.

Brown, D.J.F., Halbrendt, J.M., Robbins, R.T. & Vrain, T.C. (1993). Transmission of nepoviruses by *Xiphinema americanum*-group nematodes. *Journal of Nematology* 25, 349-354.

Castillo, P., Peña-Santiago, R. & Jiménez Millán, F. (1985). Modelos de distribución vertical de las especies de nematodos en un biotopo natural. I. *Boletín del Servicio de Defensa contra Plagas e Inspección Fitopatológica* 11, 165-172.

Castillo, P., Peña-Santiago, R., Gómez-Barcina, A. & Jiménez Millán, F. (1988). Distribución y características morfométricas de *Paractinolaimus microdentatus* (Thorne, 1939) Meyl, 1957 en la Sierra de Cazorla, Jaén (España). *Boletín de la Real Sociedad Española de Historia Natural* 84, 105-110.

CEREVKOVÁ, A. & CAGÁN, L. (2015). Effect of transgenic insect-resistant maize to the community structure of soil nematodes in two field trials. *Helminthologia* 52, 41-49. DOI: 10.1515/helmin-2015-0001

CESARZ, S., RUESS, L., JACOB, M., JACOB, A., SCHAEFER, M. & SCHEU, S. (2013). Tree species diversity *versus* tree species identity: Driving forces in structuring forest food webs as indicated by soil nematodes. *Soil Biology and Biochemistry* 62, 36-45. DOI: 10.1016/j.soilbio.2013.02.020

CHIZHOV, V.N. (1996). Phoresia of *Nygolaimoides borborophilus* (Nematoda: Dorylaimida) by scatobiotic dipterans. *Russian Journal of Nematology* 4, 185-186.

CHIZHOV, V.N., PRIDANNIKOV, M.V., PENEVA, V. & SUBBOTIN, S.A. (2014). Morphological and molecular characterisation of the Saratov population of the European dagger nematode, *Xiphinema diversicaudatum* (Nematoda: Dorylaimida), with notes on phylogeography of the species. *Nematology* 16, 847-862. DOI: 10.1163/15685411-00002813

CIANCIO, A. (1995). Phenotypic adaptations in *Pasteuria* spp. nematode parasites. *Journal of Nematology* 27, 328-338.

CIANCIO, A. & CHINAPPEN, M. (1987). *Paralongidorus buchae* host for *Catenaria anguillulae*. *Nematologia Mediterranea* 15, 391-394.

CIANCIO, A., LOGRIECO, A. & LAMBERTI, F. (1986). Parasitism of *Xiphinema diversicaudatum* by the fungus *Hirsutella rhossiliensis*. *Nematologia Mediterranea* 14, 187-192.

COHN, E. (1969). The occurrence and distribution of species of *Xiphinema* and *Longidorus* in Israel. *Nematologica* 15, 179-182. DOI: 10.1163/187529269X00614

COLL, P., LE CADRE, E. & VILLENAVE, C. (2012). How are nematode communities affected during a conversion from conventional to organic farming in southern French vineyards? *Nematology* 14, 665-676. DOI: 10.1163/156854112X624195

COOMANS, A. (1985). A phylogenetic approach to the classification of the Longidoridae (Nematoda: Dorylaimida). *Agriculture, Ecosystems and Environment* 12, 335-354. DOI: 10.1016/0167-8809(85)90006-4

COOMANS, A. & WILLEMS, A. (1998). What are symbiotic bacteria doing in the ovaria of *Xiphinema americanum*-group species? *Nematologica* 44, 323-326. DOI: 10.1163/005425998X00080

COOMANS, A., VANDEKERCKHOVE, T.T.M & CLAEYS, M. (2000). Transovarial transmission of symbionts in *Xiphinema brevicollum* (Nematoda: Longidoridae). *Nematology* 2, 443-449. DOI: 10.1163/156854100509303

COOMANS, A., HUYS, R., HEYNS, J. & LUC, M. (2001). Character analysis, phylogeny and biogeography of the genus *Xiphinema* Cobb, 1913 (Nematoda: Longidoridae). *Annales Sciences Zoologiques, Musée Royal de l'Afrique Centrale, Tervuren, Belgique* 287, 1-239.

DALMASSO, A. (1970). Influence directe de quelques facteurs écologiques sur l'activité biologique et la distribution des espèces françaises de la famille des Longidoridae (Nematoda: Dorylaimida). *Annales de Zoologie-Ecologie Animale* 2, 163-200.

DECRAEMER, W. & COOMANS, A. (1994). A compendium of our knowledge of the free-living nematofauna of ancient lakes. *Archiv für Hydrobiologie-BeiheftErgebnisse der Limnologie* 44, 173-181.

DE WAELE, D. & COOMANS, A. (1990). Occurrence and ecology of longidorid nematodes in Belgium. *Revue de Nématologie* 13, 197-202.

DU PREEZ, G., MAJDI, N., SWART, A., TRAUNSPURGER, W. & FOURIE, H. (2017). Nematodes in caves: a historical perspective on their occurrence, distribution and ecological relevance. *Nematology* 19, 627-644. DOI: 10.1163/15685411-00003068

EKSCHMITT, K. & KORTHALS, G.W. (2006). Nematodes as sentinels of heavy metals and organic toxicants in the soil. *Journal of Nematology* 38, 13-19.

FEIL, H., WESTERDAHL, B.B., SMITH, R.J. & VERDEGAAL, P. (1997). Effects of seasonal and site factors on *Xiphinema index* populations in two Californian vineyards. *Journal of Nematology* 29, 491-500.

FERRIS, H. (2010). Form and function: metabolic footprints of nematodes in the soil food web. *European Journal of Soil Biology* 46, 96-104. DOI: 10.1016/j.ejsobi.2010.01.003

FERRIS, H. & BONGERS, T. (2009). 5. Indices developed specifically for analysis of nematode assemblages. In: Wilson, M.J. & Kakouli-Duarte, T. (Eds). *Nematodes as environmental indicators*. Wallingford, UK, CAB International, pp. 124-245.

FERRIS, H. & MATUTE, M.M. (2003). Structural and functional succession in the nematode fauna of a soil food web. *Applied Soil Ecology* 23, 93-110. DOI: 10.1016/S0929-1393(03)00044-1

FERRIS, V.R. (1979). Cladistic approaches in the study of soil and plant parasitic nematodes. *American Zoologist* 19, 1195-1215. DOI: 10.1093/icb/19.4.1195

FERRIS, V.R. (1983). Phylogeny, historical biogeography and the species concept in soil nematodes. In: Stone, A.R., Platt, H.M. & Khalil, L.F. (Eds). *Concepts in nematode systematics*. The Systematic Association Special Volume No. 22. London, UK, Academic Press, pp. 143-161.

FERRIS, V.R., GOSECO, C.G. & FERRIS, J.M. (1976). Biogeography of free-living soil nematodes from the perspective of plate tectonics. *Science* 193, 508-510. DOI: 10.1126/science.193.4252.508

FERRIS, V.R., GOSECO, C.G. & FERRIS, J.M. (1980). Revisions of *Oxydirus* and *Tarjanius* n. gen. in Oxydiridae, Belondiroidea (Nematoda: Dorylaimida); and *Oxydiroides* in Prodorylaimidae, Dorylaimoidea. *Purdue University Agricultural Experiment Station Research Bulletin* 965, 1-29.

FERRIS, V.R., FERRIS, J.M. & GOSECO, C.G. (1981). Phylogenetic and biogeographic hypothesis in Leptonchidae (Nematoda: Dorylaimida) and a new classification. *Proceedings of the Helminthological Society of Washington* 48, 163-171.

FERRIS, V.R., FERRIS, J.M. & GOSECO, C.G. (1983). Revision of *Belondira* and notes on *Oxybelondira* in Belondiridae, Belondiroidea (Nematoda: Dorylaimida. *Purdue University Agricultural Experiment Station Research Bulletin* 979, 1-48.

FISCUS, D.A. & NEHER, D.A. (2002). Distinguishing sensitivity of free-living soil nematode genera to physical and chemical disturbances. *Ecological Applications* 13, 565-575. DOI: 10.2307/3060963

FLEGG, J.J.M. (1968). The occurrence and depth distribution of *Xiphinema* and *Longidorus* species in southeastern England. *Nematologica* 14, 189-196. DOI: 10.1163/187529268X00417

FRECKMAN, D.W., KAPLAN, D.T. & VAN GUNDY, S.D. (1977). A comparison of techniques for extraction and study of anhydrobiotic nematodes from dry soils. *Journal of Nematology* 9, 176-181.

GANTAIT, V.V., BHATTACHARYA, T. & CHATTERJEE, A. (2010). Seasonal fluctuation of dorylaimid population in banana plantation of West Bengal, India. *Indian Journal of Nematology* 40, 184-188.

GANTAIT, V.V., BHATTACHARYA, T. & CHATTERJEE, A. (2011a). Temporal variation in abundance of nematodes in a banana plantation of West Bengal, India. *Nematologia Mediterranea* 41, 79-82.

GANTAIT, V.V., BHATTACHARYA, T. & CHATTERJEE, A. (2011b). Studies on feeding guilds of soil and plant parasitic nematodes in banana plantation of West Bengal, India. *Indian Journal of Nematology* 39, 9-15.

GEORGIEVA, S., CHRISTENSEN, S. & SETVNBAK, K. (2005). Nematode succession and microfauna-microorganism interactions during root residue decomposition. *Soil Biology and Biochemistry* 2005, 1763-1774. DOI: 10.1016/j.soilbio.2005.02.010

GRIFFIN, G.D. & DARLING, H.M. (1964). An ecological study of *Xiphinema americanum* Cobb in an ornamental spruce nursery. *Nematologica* 10, 471-479. DOI: 10.1163/187529264X00493

HÁNĚL, L. (2001). Succession of soil nematodes in pine forests on coal-mining sands near Cottbus, Germany. *Applied Soil Ecology* 16, 23-34. DOI: 10.1016/S0929-1393(00)00101-3

HÁNĚL, L. (2003). Recovery of soil nematode populations from cropping stress by natural secondary succession to meadow land. *Applied Soil Ecology* 22, 255-270. DOI: 10.1016/S0929-1393(02)00152-X

HÁNĚL, L. (2008). Nematode assemblages indicate soil restoration on colliery spoils afforested by planting different tree species and by natural succession. *Applied Soil Ecology* 40, 86-99. DOI: 10.1016/j.apsoil.2008.03.007

HÁNĚL, L. (2010). An outline of soil nematode succession on abandoned fields in South Bohemia. *Applied Soil Ecology* 46, 355-371. DOI: 10.1016/j.apsoil.2010.10.005

HARRISON, B.D. & WINSLOW, R.D. (1961). Laboratory and field studies on the relation of arabis mosaic virus to its nematode vector *Xiphinema diversicaudatum* (Micoletzky). *Annals of Applied Biology* 49, 621-633. DOI: 10.1111/j.1744-7348.1961.tb03659.x

HARRISON, B.D., ROBERTSON, W.M. & TAYLOR, C.E. (1974). Specificity of retention and transmission of viruses by nematodes. *Journal of Nematology* 6, 155-164.

HU, C. & QI, Y. (2010). Effect of compost and chemical fertilizer on soil nematode community in a Chinese maize field. *European Journal of Soil Biology* 46, 230-236.

JAFFEE, B.A. (1986). Parasitism of *Xiphinema rivesi* and *X. americanum* by zoosporic fungi. *Journal of Nematology* 18, 87-93.

KRNJAIC, D.J. (1971). *Enchodorella deliblatica* n. sp. (Nematoda: Dorylaimida), description and distribution. *Zastita Bilja* 22, 397-408.

KRNJAIC, D.J. & LOOF, P.A.A. (1975). *Carcharolaimus banaticus* n. sp. (Nematoda: Discolaimidae) and its ecology. *Nematologia Mediterranea* 3, 153-161.

LEE, D.L. (1961). Two new species of cryptobiotic (anabiotic) freshwater nematodes, *Actinolaimus hintoni* sp. nov. and *Dorylaimus keilini* sp. nov. (Dorylaimidae). *Parasitology* 51, 237-240. DOI: 10.1017/S0031182000068621

LIÉBANAS, G., PEÑA-SANTIAGO, R., REAL, R. & MÁRQUEZ, A.L. (2002). Spatial distribution of dorylaimid and mononchid nematodes from southeast Iberian Peninsula: chorological relationships among species. *Journal of Nematology* 34, 390-395.

LIÉBANAS, G., GUERRERO, P., MARTÍN-GARCÍA, J.M. & PEÑA-SANTIAGO, R. (2004). Spatial distribution of dorylaimid and mononchid nematodes from southeast Iberian Peninsula: environmental characterization of chorotypes. *Journal of Nematology* 36, 114-122.

MACGUIDWIN, A.E. (1989). Abundance and vertical distribution of *Longidorus breviannulatus* associated with corn and potato. *Journal of Nematology* 21, 404-408.

MALIK, Z. & JAIRAJPURI, M.S. (1984). Effect of pH and some mineral salts and fatty acids on survival of *Xiphinema americanum*. *Nematologia Mediterranea* 12, 73-79.

MARSHALL, B., BOAG, B., MCNICOL, J.W. & NEILSON, R. (1998). A comparison of the spatial distributions of three plant-parasitic nematode species at three different scales. *Nematologica* 44, 303-320. DOI: 10.1163/005425998X00062

McNamara, D.G. (1980). The survival of *Xiphinema diversicaudatum* in plant-free soil. *Nematologica* 26, 170-181. DOI: 10.1163/187529280X00062

McSorley, R. (1997). Relationship of crop and rainfall to soil nematode community structure in perennial agroecosystems. *Applied Soil Ecology* 6, 147-159. DOI: 10.1016/S0929-1393(97)00001-2

McSorley, R. (2012). Ecology of the dorylaimid omnivore genera *Aporcelaimellus, Eudorylaimus* and *Mesodorylaimus. Nematology* 14, 645-663. DOI: 10.1163/156854112X651168

McSorley, R. (2013). Effect of forest and field habitat features on abundance of genera in nematode assemblages. *Nematology* 15, 947-956. DOI: 10.1163/15685411-00002733

McSorley, R. & Frederick, J.J. (1999). Nematode population fluctuations during decomposition of specific organic amendments. *Journal of Nematology* 31, 37-44.

Morone de Lucia, M.R. & Grimaldi de Zio, S. (1973). Presenza di microsporidi in gonadi di *Xiphinema mediterraneum. Nematologia Mediterranea* 1, 66-68.

Murialdo, S.E., Fuertes, M.G., González, J.F. & Chaves, E. (2002). Nematodes as indicators of wetland pollution. *Journal of Environmental Biology* 23, 423-428.

Murphy, W.S., Brodie, B.B. & Good, J.M. (1974). Population dynamics of plant nematodes in cultivated soil: effects of combinations of cropping systems and nematicides. *Journal of Nematology* 6, 103-107.

Nagy. P. (1999). Effects of an artificial metal pollution on nematode assemblage of a calcareous loamy chernozem soil. *Plant and Soil* 212, 35-43. DOI: 10.1023/A:1004657924496

Nagy, P., Bakonyi, G. & Jenser, G. (1998). Observations on the vertical distribution of *Xiphinema vuittenezi* (Longidoridae, Nematoda) in an apricot orchard in Hungary. *Nematologia Mediterranea* 26, 267-270.

Nath, R.C., Mukherjee, B. & Dasgutpa, M.K. (1998). Population dynamics of plant parasitic nematodes in a pineapple plantation of Tripura, India. *International Journal of Nematology* 8, 185-190.

Navas, A. & Arias, M. (1986). On the distribution and ecology of *Xiphinema index* and *X. italiae* in Spain. *Nematologia Mediterranea* 14, 207-215.

Navas, A., Bello, A. & Arias, M. (1988). Ecology and potential distribution of *Xiphinema diversicaudatum* and *X. pachtaicum* (Nematoda: Longidoridae) in continental Spain. *Nematologica* 34, 314-330. DOI: 10.1163/002825988X00170

Navas, A., Andrés, M.F. & Arias, M. (1990). Biogeography of Longidoridae in the Euromediterranean area. *Nematologia Mediterranea* 18, 103-112.

NAVAS, A., BALDWIN, J.G., BARRIOS, L. & NOMBELA, G. (1993). Phylogeny and biogeography of *Longidorus* (Nematoda: Longidoridae) in Euromediterranea. *Nematologia Mediterranea* 21, 71-88.

NEHER, D. & WEICHT, T.R. (2013). Nematode genera in forest soil respond differentially to elevated CO_2. *Journal of Nematology* 45, 214-222.

NGUYEN, V.C., VILLATE, L., GUTIÉRREZ-GUTIÉRREZ, C., CASTILLO, P., VAN GHELDER, C., PLANTARD, O. & ESMENJAUD, D. (2019). Phylogeography of the soil-borne vector nematode *Xiphinema index* highly suggests Eastern origin and dissemination with domesticated grapevine. *Scientific Reports* 9, 7313, 1-11. DOI: 10.1038/s41598-019-43812-4

NIBLACK, T.L. & BERNARD, E.C. (1985). Plant-parasitic nematode communities in dogwood, maple, and peach nurseries in Tennessee. *Journal of Nematology* 17, 132-139.

NORTON, D.C. & HOFFMANN, J.K. (1974). Distribution of selected plant parasitic nematodes relative to vegetation and edaphic factors. *Journal of Nematology* 6, 81-86.

NORTON, D.C., FREDERICK, L.R., PONCHILLIA, P.E. & NYHAN, J.W. (1971). Correlations of nematodes and soil properties in soybean fields. *Journal of Nematology* 3, 154-163.

PALOMARES-RIUS, J.E., ARCHIDONA-YUSTE, A., CANTALAPIEDRA-NAVARRETE, C., PRIETO, P. & CASTILLO, P. (2016). Molecular diversity of bacterial endosymbionts associated with dagger nematodes of the genus *Xiphinema* (Nematoda: Longidoridae) reveals a high degree of phylogenetic congruence with their host. *Molecular Ecology* 25, 6225-6247. DOI: 10.1111/mec.13904

PHILIS, J. (1996). The life cycle of *Xiphinema index* in Cyprus. *Nematologia Mediterranea* 24, 215-218.

PITCHER, R.S. (1975). Factors influencing the movement of nematodes in soil. In: Lamberti, F., Taylor, C.E. & Seinhorst, J.W. (Eds). *Nematode vectors of plant viruses*. London and New York, Plenum Press and NATO Scientific Affairs Division, pp. 389-407.

PLOEG. A.T. (1998). Horizontal and vertical distribution of *Longidorus africanus* in a Bermudagrass field in the Imperial Valley, California. *Journal of Nematology* 30(Suppl.), 592-598.

POPOVICI, I. (1995). Specific and trophic diversity of soil nematodes in forest ecosystems from the Zarand Mountains. *Travaux de l'Institut de Spéologie 'Emile Racovitza'* 34, 131-140.

POPOVICI, I. (1998). Structure of nematode communities in mountain grasslands from Romania. In: de Goede, R.G.M. & Bongers, T. (Eds). *Nematode communities of northern temperate grassland ecosystems*. Giessen, Germany, Focus Verlag, pp. 221-240.

PORAZINSKA, D.I., DUNCAN, L.W., MCSORLEY, R. & GRAHAM, J.H. (1999). Nematode communities as indicators of status and processes of a soil ecosystem influenced by agricultural management practices. *Applied Soil Ecology* 13, 69-86. DOI: 10.1016/S0929-1393(99)00018-9

PORAZINSKA, D.I., MCSORLEY, R., DUNCAN, L.W., GRAHAM, J.H., WHEATON, T.A. & PARSONS, L.R. (1998). Nematode community composition under various irrigation schemes in a citrus soil ecosystem. *Journal of Nematology* 30, 170-178.

POSTMA-BLAAUW, M.B., DE GOEDE, R.G.M., BLOEM, J., FABER, J.H. & BRUSSAARD, L. (2012). Agricultural intensification and de-intensification differentially affect taxonomic diversity of predatory mites, earthworms, enchytraeids, nematodes and bacteria. *Applied Soil Ecology* 57, 32-49. DOI: 10.1016/j.apsoil.2012.02.011

PROTA, U. & GARAU, R. (1973). Indagini sulla biologia di *Xiphinema index* Thorne & Allen in vigneti sardi. *Nematologia Mediterranea* 1, 36-54.

ROCCUZZO, G. & CIANCIO, A. (1991). Notes on nematodes found in irrigation water in southern Italy. *Nematologia Mediterranea* 19, 105-108.

RUESS, L., SCHMIDT, I.K., MICHELSEN, A. & JONASSON, S. (2002). Responses of nematode species composition to factorial addition of carbon, fertiliser, bactericide and fungicide at two sub-arctic sites. *Nematology* 4, 527-539. DOI: 10.1163/156854102760290509

SAYRE, R.M. & STARR, M.P. (1988). Bacterial diseases and antagonisms of nematodes. In: Poinar Jr, G.O. & Jansson, H.B. (Eds). *Diseases of nematodes II*. Boca Raton, FL, USA, CRC Press, pp. 69-102. DOI: 10.1201/9781351071475-5

STIRLING, G.R. (1985). Observations on the biology and ecology of *Paralongidorus australis* on rice in North Queensland. *Nematologica* 31, 218-228. DOI: 10.1163/187529285X00283

STURHAN, D. (1988). New host and geographical records of nematode-parasitic bacteria of the *Pasteuria penetrans* group. *Nematologica* 34, 350-356. DOI: 10.1163/002825988X00198

SULTAN, S.A. & FERRIS, H. (1991). The effect of soil moisture and soil particle size on the survival and population increase of *Xiphinema index*. *Revue de Nématologie* 14, 345-351.

TAHSEEN, Q., CLARK, I.M. & KERRY, B.R. (2003). *Vorticella* sp. infestations in nematodes: a report. *Nematologia Mediterranea* 31, 195-196.

TAYLOR, C.E. & BROWN, D.J.F. (1997). *Nematode vectors of plant viruses*. Wallingford, UK, CAB International.

TAYLOR, C.E. & ROBERTSON, W.M. (1975). Acquisition, retention and transmission of viruses by nematodes. In: Lamberti, F., Taylor, C.E. & Seinhorst, J.W. (Eds). *Nematode vectors of plant viruses*. London and New York, Plenum Press and NATO Scientific Affairs Division, pp. 253-276.

TENUTA, M. & FERRIS, H. (2004). Sensitivity of nematode life-history groups to ions and osmotic tensions of nitrogenous solutions. *Journal of Nematology* 36, 85-94.

THOMAS, P.R. (1981). Migration of *Longidorus elongatus, Xiphinema diversicaudatum* and *Ditylenchus dipsaci* in soil. *Nematologia Mediterranea* 9, 75-81.

THORNE, G. (1930). Predaceous nemas of the genus *Nygolaimus* and a new genus *Sectonema*. *Journal of Agricultural Research* 41, 445-466.

TODD, T.C., POWERS, T.O. & MULLIN, P.G. (2006). Sentinel nematodes of land-use change and restoration in tallgrass prairie. *Journal of Nematology* 38, 20-27.

TOPHAM, P.B. & ALPHEY, T.J.W. (1985). Faunistic analysis of longidorid nematodes in Europe. *Journal of Biogeography* 12, 165-174. DOI: 10.2307/2844839

VANDEKERCKHOVE, T.T.M., COOMANS, A., CORNELIS, K., BAERT, P. & GILLIS, M. (2002). Use of the Verrucomicrobia-specific robe EUB338-III and fluorescent *in situ* hybridization for detection of "*Candidatus Xiphinematobacter*" cells in nematode hosts. *Applied and Environmental Microbiology* 68, 3121-3125. DOI: 10.1128/AEM.68.6.3121-3125.2002

VELASCO-CASTRILLÓN, A., GIBSON, J.A.E. & STEVENS, M.I. (2014). A review of current Antarctic limno-terrestrial microfauna. *Polar Biology* 37, 1517-1531. DOI: 10.1007/s00300-014-1544-4

VIDAKOVIĆ, J. & BOGUT, I. (2006). Aquatic macrophytes as a habitat for free-living nematodes. *Nematology* 8, 691-701. DOI: 10.1163/156854106778877956

VILLENAVE, C., CADET, P., PLANCHON, O., ESTÈVE, M. & LAPETITE, J.M. (2003). Transport of free-living nematodes by runoff water in a Sudano-Sahelian area. *Applied Soil Ecology* 23, 85-91. DOI: 10.1016/S0929-1393(02)00176-2

VINCIGUERRA, M.T. (1988). A new classification of Actinolaimoidea Thorne, 1939 using a cladistic approach. *Nematologica* 33(1987), 251-277. DOI: 10.1163/187529287X00399

WALIULLAH, M.I.S. (1984). Nematodes in irrigation water. *Nematologia Mediterranea* 12, 243-245.

WALIULLAH, M.I.S. (1989). Nematodes in irrigation canals of the Kashmir Valley, India. *Nematologia Mediterranea* 17, 55-56.

WALL, D.H. & VIRGINIA, R.A. (1999). Controls on biodiversity: insights from extreme environments. *Applied Soil Ecology* 13, 137-150. DOI: 10.1016/S0929-1393(99)00029-3

WALTER, D.E. & IKONEN, E.K. (1989). Species, guilds, and functional groups: taxonomy and behavior in nematophagous arthropods. *Journal of Nematology* 21, 315-327.

WANG, K.-H., MCSORLEY, R. & KOKALIS-BURELLE, N. (2006). Effects of cover cropping, solarization, and soil fumigation on nematode communities. *Plant and Soil* 286, 229-243. DOI: 10.1007/s11104-006-9040-4

WASILEWSKA, L. (1997). Soil invertebrates as bioindicators, with special reference to soil-inhabiting nematodes. *Russian Journal of Nematology* 5, 113-126.

WEISCHER, B. (1975). Ecology of *Xiphinema* and *Longidorus*. In: Lamberti, F., Taylor, C.E. & Seinhorst, J.W. (Eds). *Nematode vectors of plant viruses*. London and New York, Plenum Press and NATO Scientific Affairs Division, pp. 291-307.

WEISCHER, B. & ALMEIDA, M.T.M. (1995). Ecology of longidorid nematodes. *Russian Journal of Nematology* 3, 9-21.

WIT, R. DE & BOUVIER, T. (2006). *'Everything is everywhere*, but, *the environment selects'*; what did Baas Becking and Beijerinck really say? *Environmental Microbiology* 8, 755-758. DOI: 10.1111/j.1462-2920.2006.01017.x

YEATES, G.W. (1967). Studies on nematodes from dune sands. 9. Quantitative comparison of the nematode faunas of six localities. *New Zealand Journal of Science* 10, 927-948.

YEATES, G.W. (1973). Nematoda of a Danish beech forest. II. Production estimates. *Oikos* 24, 179-185. DOI: 10.2307/3543873

YEATES, G.W. (2003). Nematodes as soil indicators: functional and biodiversity aspects. *Biology and Fertility of Soils* 37, 199-210. DOI: 10.1007/s00374-003-0586-5

ZHAO, J. & NEHER, D.A. (2013). Soil nematode genera that predict specific types of disturbance. *Applied Soil Ecology* 64, 135-141. DOI: 10.1016/j.apsoil.2012.11.008

ZOON, F.C., TROELSTRA, S.R. & MAAS, P.W.T. (1993). Ecology of the plant-feeding nematode fauna associated with sea buckthorn (*Hippophaë rhamnoides* L. ssp. *rhamnoides*) in different stages of dune succession. *Fundamental and Applied Nematology* 16, 247-258.

ZULLINI, A. (1970). I nematodi muscicoli della Val Zebru (Parco Nazionale dello Stelvio). *Rendiconti: Istituto Lombardo Accademia di Scienze e Lettere* 104, 88-137.

ZULLINI, A. (2014). Is a biogeography of freshwater nematodes possible? *Nematology* 16, 1-8. DOI: 10.1163/15685411-00002779

ZULLINI, A. (2018). Cosmopolitism and endemism in free-living nematodes. *Biogeographica* 33, 33-39. DOI: 10.21426/B633034658

ZULLINI, A., GATTI, F. & AMBROSINI, R. (2011). Microhabitat preferences in springs, as shown by a survey of nematode communities of Trentino (south-eastern Alps, Italy). *Journal of Limnology* 70(Supplement), 93-105. DOI: 10.4081/jlimnol.2011.s1.93

Nematology Monographs & Perspectives, 2021, Vol. 13, 231-267

Chapter 14

Diversity

Historical outline

Our knowledge of dorylaims started nearly two centuries ago. Since then, the history of the group, in particular the study of its diversity, can be divided into five main stages.

THE ORIGINS (1845-1920)

Dujardin (1845) described the first dorylaimid species, *Dorylaimus stagnalis*, from which the ordinal name derived. Twenty years later, Bastian (1865) emended the diagnosis of the genus, added new species and transferred two more species to the genus. Bütschli (1873) provided re-descriptions of Bastian's species. A family rank for the group was proposed by de Man (1876), with *Dorylaimus* remaining as its only genus. The family Dorylaimidae was already characterised on the basis of a morphological pattern very similar to that accepted nowadays for the order Dorylaimida: axial odontostyle with distinct aperture, bipartite and bottle-like pharynx, distinct prerectum, ventrad arcuate spicules, lateral guiding pieces and ventromedian supplements. Many genera were subsequently classified under Dorylaimidae by several pioneer authors (de Man, 1880; Örley, 1880; Cobb, 1913, 1920), and the definition of the group was substantially modified. At the end of the 19th century, two genera and 84 species had already been described, and their number significantly increased to 16 genera and 208 species in the course of the next 30 years.

THE PRODIGIOUS 1930S DECADE

Filipjev (1929, 1934) introduced significant changes in the classification of Dorylaimidae, with the recognition of four subfamilies, namely Alaiminae, Dorylaiminae, Ironinae and Tylencholaiminae. Nevertheless, the most relevant advance, with a new approach, was provided by Gerald Thorne in a series of papers that appeared during the 1930s. In 1934, he raised the rank of the group, with proposal of the superfamily Dorylaimoidea, and recognised the families Dorylaimidae and Alaimidae. One year later (Thorne, 1935) he proposed the new subfamilies Longidorinae and Nygolaiminae under Dorylaimidae along with the new families Leptonchidae and Diphtherophoridae. The taxonomy of dorylaims was exhaustively revised and consolidated in two classical monographs (Thorne & Swanger, 1936; Thorne, 1939), which are still keystones in the study of this taxon due to the high number of new taxa described and the good quality of their illustrations. At the end of this decade the catalogue of dorylaims had increased substantially to include 38 genera and 449 species.

The order Dorylaimida setting

Pearse first (Pearse, 1936) raised the rank of dorylaims to subordinal rank (Dorylaimina) and later (Pearse, 1942) to ordinal rank (Dorylaimida). Nevertheless, the last proposal did not gain general acceptance (Thorne, 1939; Clark, 1961) until the revision of the classical textbook *Soil and freshwater nematodes* by J.B. Goodey (1963). The concept of the group was, at that time, quite different from the currently accepted scheme since, for example, the alaims, including trichodorid forms, were considered to be a suborder, and the mononchs a superfamily of Dorylaimida (Table 14.1).

The golden age of the exploration of dorylaimid diversity (1960-1990)

After the long pause caused by the Second World War, and during the second half of the 20th century, a few laboratories once again took up the study of various aspects of dorylaimid diversity, utilising new approaches and the development of new techniques. The inventory of existing genera and species experienced an incredible expansion, reaching 284 genera and 2264 species by the end of the period, through the efforts of reputed specialists such as: I. Andrássy, A. Coomans, V. Ferris, J. Heyns, M.S. Jairajpuri, P.A.A. Loof, M.R. Siddiqi, G. Thorne, M.T. Vinciguerra, and A. Zullini, together with their collaborators.

The morphology of the group was beautifully described, illustrated and explained by the Belgian nematologists of the Ghent school, headed

Table 14.1. *Classification of Dorylaimida (families) as proposed by several authors throughout the last half century. Suborders are in bold, superfamilies in small caps. Taxa later regarded as non-dorylaimid groups are in square brackets.*

Goodey (1963)	Andrássy (1976)	Jairajpuri & Ahmad (1992)	De Ley & Blaxter (2004)	Andrássy (2009)	Peña-Santiago (2014)
Dorylaimina	**Dorylaimina**	**Nygolaimina**	**Nygolaimina**	**Nygolaimina**	**Nygolaimina**
DORYLAIMOIDEA	ENCHOLAIMOIDEA	NYGOLAIMOIDEA	NYGOLAIMOIDEA	NYGOLAIMOIDEA	Aetholaimidae
Dorylaimidae	Encholaimidae	Aetholaimidae	Aetholaimidae	Aetholaimidae	Nygellidae
Opailaimidae	NYGOLAIMOIDEA	Nygellidae	Nygellidae	Nygellidae	Nygolaimellidae
Belondiridae	Nygolaimidae	Nygolaimellidae	Nygolaimellidae	Nygolaimidae	Nygolaimidae
Nygolaimidae	Aetholaimidae	Nygolaimidae	Nygolaimidae	**Dorylaimina**	**Dorylaimina**
[Campydoridae]	DORYLAIMOIDEA	**Dorylaimina**	**Dorylaimina**	DORYLAIMOIDEA	Actinolaimidae
[MONONCHOIDEA]	Prodorylaimidae	DORYLAIMOIDEA	DORYLAIMOIDEA	Actinolaimidae	Aporcelaimidae
[Mononchidae]	Dorylaimidae	Aporcelaimidae	Actinolaimidae	Aporcelaimidae	Aulolaimoididae
[Bathyodontidae]	Thornenematidae	Dorylaimidae	Aporcelaimidae	Dorylaimidae	Belondiridae
[Alaimina]	Qudsianematidae	Nordiidae	Dorylaimidae	Nordiidae	Dorylaimidae
[ALAIMOIDEA]	Aporcelaimidae	Qudsianematidae	Longidoridae	Paraxonchiidae	Leptonchidae
[Alaimidae]	Nordiidae	ACTINOLAIMOIDEA	Nordiidae	Qudsianematidae	Longidoridae
[DIPHTHEROPHOROIDEA]	Longidoridae	Actinolaimidae	Qudsianematidae	Thornenematidae	Mydonomidae
[Diphtherophoridae]	Dorylaimoididae	Carcharolaimidae	BELONDIROIDEA	Thorniidae	Nordiidae
[Trichodoridae]	Crateronematidae	LONGIDOROIDEA	Belondiridae	LONGIDOROIDEA	Qudsianematidae
	Thorniidae	Longidoridae	TYLENCHOLAIMOIDEA	Longidoridae	Thornenematidae
	BELONDIROIDEA	Xiphinematidae	Aulolaimoididae	BELONDIROIDEA	Thorniidae
	Oxydiridae	BELONDIROIDEA	Leptonchidae	Belondiridae	Tylencholaimellidae
	Swangeriidae	Belondiridae	Mydonomidae	Dorylaimellidae	Tylencholaimidae
	Roqueidae	TYLENCHOLAIMOIDEA	Tylencholaimidae	Swangeriidae	
	Belondiridae	Aulolaimoididae		TYLENCHOLAIMOIDEA	
	Dorylaimellidae	Leptonchidae		Aulolaimoididae	

Table 14.1. *(Continued.)*

Goodey (1963)	Andrássy (1976)	Jairajpuri & Ahmad (1992)	De Ley & Blaxter (2004)	Andrássy (2009)	Peña-Santiago (2014)
	ACTINOLAIMOIDEA	Mydonomidae		Encholaimidae	
	Trachypleurosidae	Tylencholaimidae		Leptonchidae	
	Actinolaimidae	**[Campydorina]**		Mydonomidae	
	Brittonematidae	[CAMPYDOROIDEA]		Tylencholaimellidae	
	Carcharolaimidae	[Campydoridae]		Tylencholaimidae	
	LEPTONCHOIDEA				
	Tylencholaimidae				
	Leptonchidae				
	Tylencholaimellidae				
	Belonenchidae				
	Aulolaimoididae				
	[Campydoridae]				
	[Mononchina]				
	[Diphtherophorina]				
	[Mermithina]				

by A. Coomans, during the 1960s and 1970s. The taxonomy of dorylaims was addressed in a myriad of papers, deserving special mention the extraordinary contribution of I. Andrássy (see the compilation by Peña-Santiago, 2013) as well other relevant publications by Heyns (1965, 1968), Siddiqi (1969), Thorne (1964, 1967), Goseco *et al.* (1974), Coomans & Nair (1975) and Carbonell & Coomans (1986), among many others, ending with the monograph *Dorylaimida* by Jairajpuri & Ahmad (1992). Phylogeny and systematics received less attention in this period, but Andrássy (1976) tried to elucidate the evolutionary relationships between the major nematode groups, although his ideas did not gain general acceptance (see Coomans, 1977). Several contributions (Coomans, 1985; Vinciguerra, 1987) used the cladistic approach to try to understand these aspects of selected dorylaimid taxa.

A NEW ERA FOR AN INTEGRATIVE APPROACH

The study of dorylaimid diversity has continued throughout the three decades since 1990 with a traditional (morphological) perspective, and the number of new genera and species has continued to grow. The paper by Blaxter *et al.* (1998) opened a new era for the characterisation of nematode diversity and the systematics of the phylum with an integrative approach, *i.e.*, based on a combination of data provided by different tools, especially those of a morphological and molecular nature (De Ley & Blaxter, 2002, 2004). Clearly, dorylaims were not exempt from this approach.

The first findings, derived from the analysis of both SSU and LSU rDNA sequences, demonstrated that dorylaims (order Dorylaimida) and their two major subgroups (suborders Dorylaimina and Nygolaimina) were natural (monophyletic) taxa (Mullin *et al.*, 2005), but also that their superfamilies and most families were not monophyletic (Holterman *et al.*, 2008). This meant that the traditional classification, based on morphological features, was not at all satisfactory, and that further studies would be necessary to overcome the current viewpoint, with at least two lines of action: first, the re-interpretation of morphological data, and, second, the gathering of new molecular information, either through sequencing new taxa and/or new genes.

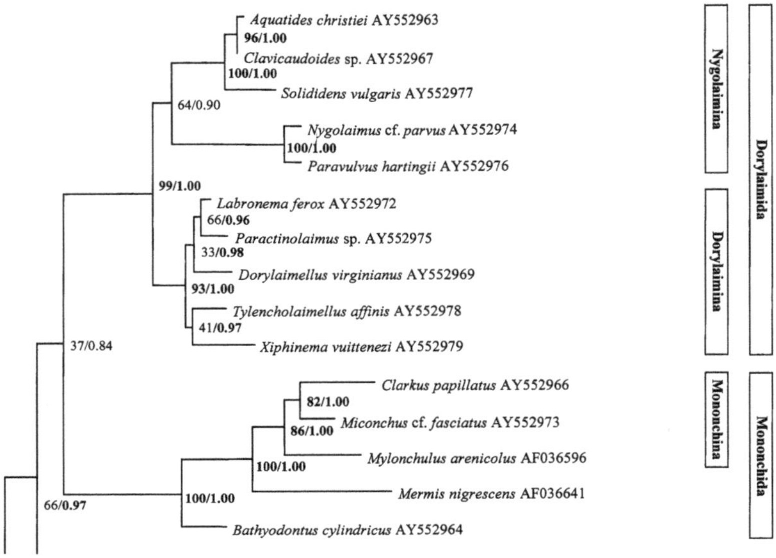

Fig. 14.1. *Molecular tree of dorylaimid taxa showing the primary division into two suborders, Nygolaimina and Dorylaimina, using mononchs as outgroup. (After Mullin* et al. *(2005),* Nematology *7, p. 68, Fig. 2.)*

The internal phylogeny of Dorylaimida

As already mentioned, and as currently conceived, the dorylaims constitute a natural (monophyletic) group. From a cladistic, morphological perspective, it is defined by having a unique combination of characters (protrusible stomatal structure, two-sectioned, bottle-like pharynx, prerectum, pre-cloacal pair of genital papillae or supplements, and the absence of caudal glands), with at least two of these characters (prerectum, pre-cloacal pair of genital papillae) considered to be recognisable and exclusive autapomorphic conditions. Also, dorylaimid sperm cells show some exclusive characters (Yushin, 2010). Molecular analyses have also confirmed (Blaxter *et al.*, 1998; Holterman *et al.*, 2008) the monophyly of the order.

Both morphological and molecular data also support the primary branching of the dorylaimid evolutionary tree into two main subdivisions (suborders), namely Dorylaimina and Nygolaimina (Mullin *et al.*, 2005; Fig. 14.1). The nature of the protrusible stomatal structure, either an axial odontostyle or a mural tooth, is the key morphological feature to separate these suborders, the former being considered as the plesiomorphic condition with the latter being apomorphic. Nygolaims, the representatives of Nygolaimina, are also characterised by the presence of three large cardiac cells that, very occasionally and never with comparable development, also exist in dorylaims *sensu stricto* (Dorylaimina). Dorylaimina is by far the largest and most speciose suborder, including 95.5% of the valid species and 95.1% of the valid genera (*cf.* Andrássy, 2009).

The secondary branching of the Dorylaimida evolutionary tree is, however, more controversial. On the one hand, the traditional (morphological) classification (Jairajpuri & Ahmad, 1992; Andrássy, 2009) split the two suborders into a variable number of superfamilies and families (Table 14.1). Thus, the Nygolaimina system consisted of three or four families, grouped under only one superfamily, the Nygolaimoidea, whereas the number of superfamilies and families in Dorylaimina was more variable (3-5, 13-18, respectively). On the other hand, the most comprehensive molecular analyses of LSU-rDNA gene sequences[1] (Holterman *et al.*, 2008) split the suborder Dorylaimina into 12 clades, confirmed the monophyly of only two superfamilies (Longidoroidea and Actinolaimoidea), and did not provide strong support to maintain the other superfamilies (Belondiroidea, Dorylaimoidea and Tylencholaimoidea) or several families as being monophyletic taxa. With a more integrative approach, De Ley & Blaxter (2004) reduced the number of superfamilies and families, and Peña-Santiago (2014) presented a more simplified system leaving aside the superfamilies.

Evolutionary relationships of Dorylaimida with other nematode taxa

Nematode internal phylogeny is an intricate subject and a continual matter of controversy. Traditionally, it seemed nearly impossible to have a common systematic system covering free-living marine, animal-parasitic and continental (free-living and plant-parasitic) forms, the general classifications being often visibly biased. Fortunately, modern techniques, especially sequencing, have provided powerful tools to overcome the situation and made it easier to accomplish an integrative

[1] The analyses of SSU-rDNA gene sequences did not provide a good resolution of the Dorylaimida tree (Holterman *et al.*, 2008; van Megen *et al.*, 2009).

approach. Thus, the evolutionary relationships of dorylaims with other nematode taxa have been significantly modified during the last five decades (summarised in Table 14.2) according to the changes introduced in the definition of the group and/or the availability of new data.

DORYLAIMS – MEMBERS OF THE SUBCLASS DORYLAIMIA

Several integrative proposals (De Ley & Blaxter, 2002, 2004; Medal *et al.*, 2007; Bert *et al.*, 2011; Hodda, 2011) present an evolutionary scenario for Nematoda with a primary division into two classes, Enoplea and Chromadorea, with the former being the basal group of nematodes and the sister group of the latter (Bert *et al.*, 2011). Previously, and with a morphological perspective, Malakhov (1998) also postulated that enoplid groups are probably the most primitive nematodes as several features, especially those concerning their embryological development, are different to those in all other nematode taxa. Most contributions also agree that dorylaims form part of a monophyletic group, the subclass Dorylaimia, which also includes the free-living mononchs and several animal-parasitic taxa, the subclass Enoplia being its sister group. The monophyly of Dorylaimia is supported by a few conservative molecular synapomorphies, the unique specification of the endodermal precursor in early embryogenesis, and no unequivocal adult morphological synapomorphy (Rusin *et al.*, 2003).

MONONCHS – THE CLOSEST RELATIVES OF DORYLAIMS?

There is a general consensus that, among the free-living nematodes, the mononchs (order Mononchida) are the closest group to the dorylaims. Actually, both taxa have very often been classified with a similar rank, either as superfamilies (Goodey, 1963; De Coninck, 1965), suborders (Coomans & Loof, 1970, 1986; Andrássy, 1976), orders (Jairajpuri, 1969; Maggenti, 1982; Inglis, 1983; Malakhov, 1994) or even superorders or subclasses (Hodda, 2007, 2011). Dorylaims and mononchs have in common several relevant morphological features, for example, the presence of five pharyngeal glands opening posterior to the nerve ring, and a totally comparable morphology of both the ejaculatory and rectal glands of the male reproductive system (Coomans & Loof, 1986). Nevertheless, they differ significantly not only in the nature of their stoma, the morphology of their pharynx and the arrangement of the pharyngeal gland nuclei (Coomans & Loof, 1970; Loren-

Table 14.2. *Classification schemes of Dorylaimida and their relatives throughout the last half century. Typical dorylaimid taxa are in bold. Taxa regarded as dorylamid taxa in some classifications are in square brackets. The taxa in brackets are mentioned to give an idea of the general classification of Nematoda. Other taxa that were never considered to be close to dorylaims are not included.*

Scheme	Class	Subclass	Superorder	Order	Suborder	Superfamily
De Coninck (1965)						
	Adenophorea	Enoplia		Enoplida		[Mermithoidea]
						Trichuroidea
						Dioctophymatoidea
						Muspiceoidea
				Dorylaimida	**Dorylaimina**	[Mononchoidea]
						Nygolaimoidea
						Dorylaimoidea
						[Diphtherophoroidea]
					[Alaimina]	
		(Chromadoria)				
	Secernentea					
Andrássy (1976)						
	Nematoda	Penetrantia		Enoplida	Tripylina	Oxystominoidea
				Dorylaimida	[Mononchina]	
					Dorylaimina	**Encholaimoidea**
						Nygolaimoidea
						Dorylaimoidea
						Belondiroidea
						Actinolaimoidea
						Leptonchoidea
					[Diphtherophorina]	
					[Mermithina]	
				Trichocephallida		
				Dioctophymatida		

Table 14.2. *(Continued.)*

Scheme	Class	Subclass	Superorder	Order	Suborder	Superfamily
		(Secernentia)				
		(Torquentia)				
MAGGENTI (1982)						
	Adenophorea	Enoplia		Enoplida		
				Isolaimida		
				[Mononchida]		
				Dorylaimida	**Dorylaimina**	
					[Diphtherophorina]	
					Nygolaimina	**Nygolaimoidea**
					[Alaimina]	
				Trichocephalida		
				[Mermithida]		
				Muspiceida		
		(Chromadoria)				
	(Secernentea)	(Rhabditia)		(Ascaridida)	Dioctophymatina	
INGLIS (1983)						
	(Rhabditea)					
	Enoplea	Enoplia		Enoplida		
		Dorylaimia		**Dorylaimida**	**Dorylaimina**	
					[Alaimina]	
					[Bathyodontina]	
					[Diphtherophorina]	
				Dioctophymatida		
				[Mermithida]		
				[Mononchida]		
				Muspiceida		
				Trichurida		
	(Chromadorea)					

Table 14.2. *(Continued.)*

Scheme	Class	Subclass	Superorder	Order	Suborder	Superfamily
Malakhov (1994)						
	Nematoda	Enoplia		Enoplida		
				Marimermithida		
				[Mononchida]		
				Dorylaimida	**Dorylaimina**	
					[Diphtherophorina]	
				[Mermithida]		
				Trichocephalida		
				Dioctophymida		
		(Chromadoria)				
		(Rhabditia)				
De Ley & Blaxter (2004)						
	Enoplea	Enoplia		Enoplida	[Campydorina]	
					[Alaimina]	
				[Triplonchida]		
		Dorylaimia		**Dorylaimida**	**Dorylaimina**	**Dorylaimoidea**
						Belondiroidea
						Tylencholaimoidea
					Nygolaimina	**Nygolaimoidea**
				[Mononchida]		
				Isolaimida		
				Dioctophymatida		
				Muspiceida		
				Marimermithida		
				[Mermithida]		
				Trichinellida		
	(Chromadorea)					

Table 14.2. *(Continued.)*

Scheme	Class	Subclass	Superorder	Order	Suborder	Superfamily
HODDA (2011)						
	Enoplea	Enoplia	Enoplica	Ironida	[Campydorina]	
				[Alaimida]		
		[Triplonchia]	Triplonchica	[Diphtherophorida]		
		[Bathyodontia]	Mononchica	[Bathyodontida]		
				[Mononchida]		
				[Mermithida]	Aulolaimina	Isolaimoidea
		Dorylaimia	Dorylaimica	**Dorylaimida**	**Dorylaimina**	**Dorylaimoidea**
						Actinolaimoidea
						Longidoroidea
						Belondiroidea
						Tylencholaimoidea
					Nygolaimina	**Nygolaimoidea**
		Trichocephalia	Trichocephallica	Dioctophymatida		
				Marimermithida		
				Muspiceida		
				Trichocephalida		
	(Chromadorea)					

zen, 1994), but also in the morphology of their pharyngo-intestinal junction, and the presence/absence of prerectum and caudal glands with a spinneret (Jairajpuri, 1969). As stated by Coomans & Loof (1986), the mononchs are a heterogeneous group since they include at least two major subgroups (bathyodonts and mononchs *sensu stricto* or Bathyodontina and Mononchina, respectively), which share the cylindrical shape of their pharynx and differ from each other mainly in the nature of their stoma; moreover, in turn, the bathyodonts may be split into two separate taxa, namely the bathyodonts *sensu stricto* and the mononchuls (Bathyodontoidea and Mononchuloidea, respectively), which also differ significantly in the nature of their stoma. As derived from a cladistic analysis of the nature of the stoma and the arrangement of the pharyngeal gland nuclei, the evolution of dorylaims and mononchs from an enoplid ancestor can be hypothesised as follows: (Mononchina (Bathyodontoidea (Mononchuloidea (Dorylaimida)))) (Coomans & Van der Heiden, 1979; Fig. 14.2). Molecular analyses of SSU gene sequences, however, have repeatedly shown that mononchs are closer to the animal-parasitic mermiths than to dorylaims, the three groups forming a robust clade supported by two molecular synapomorphies and, at present, no adult morphological synapomorphy (Rusin *et al.*, 2003). Thus, mononchs *sensu lato* and mermithids have been clustered under the superorder Mononchica, the only one in the subclass Bathyodontia (Hodda, 2007, 2011; Table 14.2).

Dorylaims and other animal-parasitic Dorylaimia

Morphological and molecular evidence suggest that two other animal-parasitic orders, namely Dioctophymatida and Trichinellida, constitute a recognisable group that may be the sister group of the (Dorylaimida (Mononchida (Mermithida))) clade, and form part of Dorylaimia. Nevertheless, the precise nature of the evolutionary relationships between these five orders has not been definitively elucidated and several alternative hypotheses (Fig. 14.3) have been proposed for their phylogeny (Inglis, 1983; Malakhov, 1994; Rusin *et al.*, 2003; Yushin & Malakhov, 2004).

De Ley & Blaxter (2004) also classified two other animal-parasitic orders, Marimermithida and Muspiceida, under Dorylaimia, but the relationship of these taxa with other members of the subclass remain uncertain. Muspiceida were regarded as a link between Mermithida and

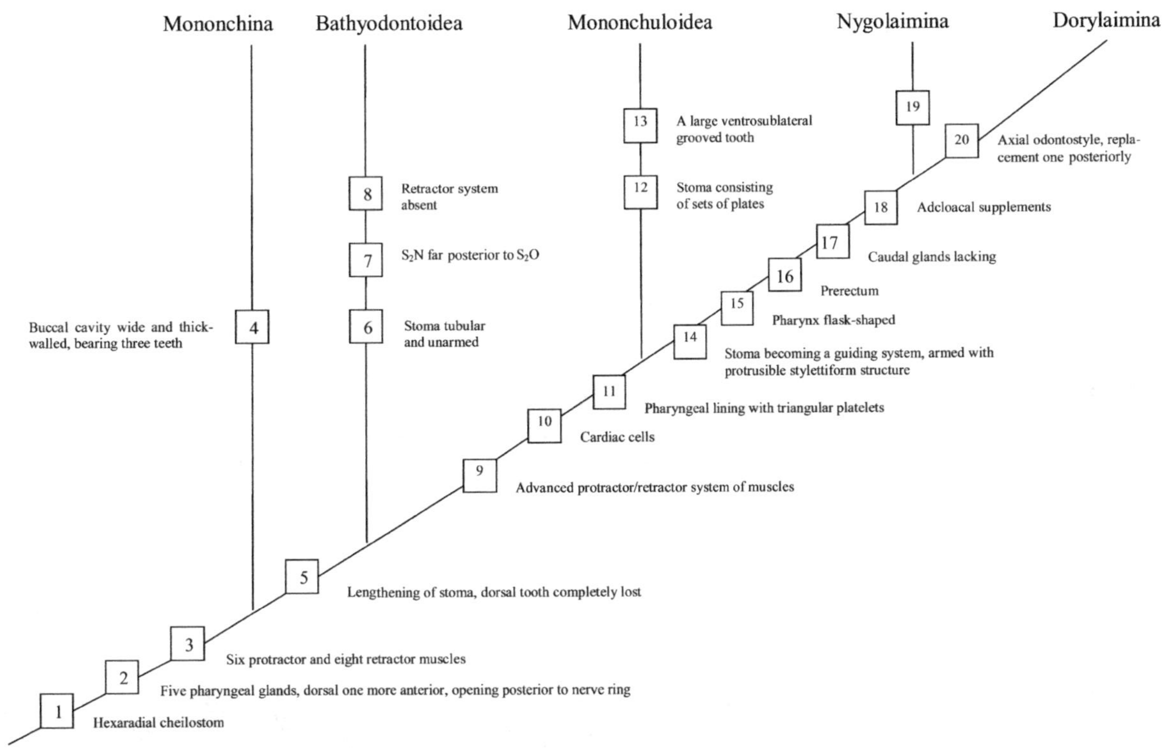

Fig. 14.2. *Cladogram of free-living suborders of the subclass Dorylaimia, with indication of apomorphic conditions. (Based on Coomans & Van der Heiden (1979),* Annales de la Société Royal Zoologique de Belgique *108, p. 9, Fig. 1.)*

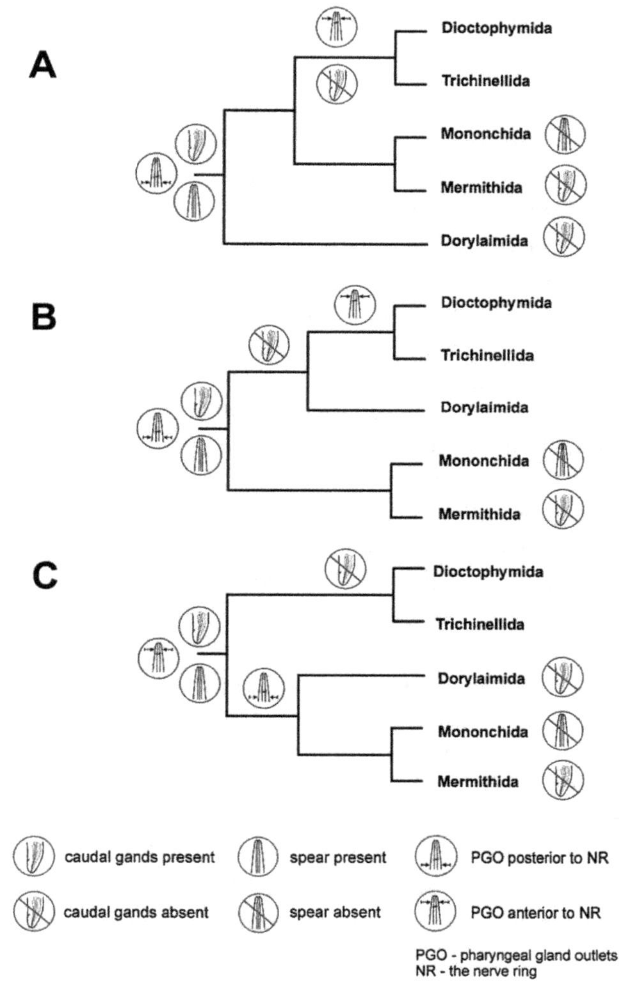

Fig. 14.3. *Three possible internal phylogenies of Dorylaimia based on morphological features. (After Rusin* et al. *(2003),* Nematology *5, p. 621, Fig. 3.)*

Trichinellida (Spratt & Nicholas, 2002), but the absence of molecular data prohibits a confirmation of that proposal (Spratt, 2014). Regarding Marimermithida, Malakhov (1994) hypothesised that this group of parasites of marine invertebrate animals might derive from (or have an early common ancestor with) ancestral enoplid forms, a proposal that might find support with molecular data (*cf.* Miljutin, 2014).

DORYLAIMIDA AND OTHER NON-DORYLAIMIA

De Ley & Blaxter (2004) also classified the order Isolaimida in the subclass Dorylaimia. The isolaims have an intricate taxonomical history, with diverse interpretations of their evolutionary relationships (summarised by Holovachov, 2014), but are very often considered to be more or less close to dorylaims. They, however, differ significantly from these in several key features, especially their tube-like, unarmed stoma, cylindrical pharynx and absence of a prerectum. Hodda (2011) regarded isolaims as a sister group of mermithids and therefore as components of the subclass Bathyodontia. Molecular data show that isolaims belong to a clade also containing Rhabditida, Plectida and other free-living groups (Medal *et al.*, 2007; Abolafia & Peña-Santiago, 2018).

Several members of the class Enoplea, now classified under the subclass Enoplia, have been traditionally regarded as more or less close to dorylaims, among others: alaims, *Campydora*, Enoplida and triplonchs (Table 14.2; taxa in square bracket).

Alaims resemble dorylaims in their two-sectioned, bottle-like pharynx with glands opening posterior to the nerve ring, but both groups differ significantly in the morphology of the stoma, which is totally unarmed in the former. Nevertheless, alaims were considered to be a member of the Dorylaimida with a subordinal rank (Dorylaimina and Alaimina; Goodey, 1963; De Coninck, 1965; Maggenti, 1982; Inglis, 1983; Lorenzen, 1994), or even as a separate order Alaimida (Siddiqi, 1983). Subsequently, molecular data demonstrated that alaims are closer to Enoplida than to Dorylaimida (Blaxter *et al.*, 1998), being regarded as a suborder of the former (De Ley & Blaxter, 2004) or a separate order (Hodda, 2011) as internal relationships with enoplid taxa are not yet definitively resolved (Medal *et al.*, 2007).

For a long time, the rare and enigmatic genus *Campydora* was considered to be a dorylaimid taxon, either as a family (Campydoridae) (Goodey, 1963; De Coninck, 1965; Andrássy, 1976; Maggenti, 1982) or a suborder (Campydorina) (Jairajpuri, 1983; Jairajpuri & Ahmad, 1992). *Campydora* displays a very special morphological pattern characterised by, among other traits, the absence of lateral chords, lips exceptionally large and hyaline, a subdorsal hollow mural tooth, pharynx ending in a basal bulb with a distinct, triquetrous chamber, excretory pore and duct well perceptible, and absence of a prerectum (Jairajpuri, 1983). This pattern does not compare to that of the Dorylaimida (*cf.* Siddiqi, 1983).

Molecular analyses have confirmed that it is closer to enoplid rather than to dorylaimid taxa (Mullin *et al.*, 2004). Thus, more recent classifications regard *Campydora* as the only genus of the suborder Campydorina, an enoplid taxon (De Ley & Blaxter, 2004; Hodda, 2011).

Enoplida and Dorylaimida were traditionally considered to be closely related taxa, but always as separate orders. Lorenzen (1994) clustered both groups along with the order Trefusiida in the subclass Enoplia, failed to find any morphological holapomorphy characterising this taxon, but distinguished Enoplida from Dorylaimida by the presence/absence of metanemes, respectively, and in the anterior or posterior opening of pharyngeal gland nuclei, also respectively, the latter a questionable diagnostic feature (*cf.* Medal *et al.*, 2007). Molecular data totally support an unquestionable separation of Enoplida and Dorylaimida, which are classified in separate subclasses (Enoplia and Dorylaimia, respectively) of the class Enoplea (De Ley & Blaxter, 2002, 2004) or even in separate classes (Enoplea and Dorylaimea; see Hodda, 2007, 2011).

Many classifications have recognised the triplonchs, especially the diphtherophorids, as members of the order Dorylaimida, either as a superfamily (Diphtherophoroidea) (Goodey, 1963; De Coninck, 1965) or a suborder (Diphtherophorina) (Coomans & Loof, 1970; Andrássy, 1976; Maggenti, 1982; Inglis, 1983; Lorenzen, 1994; Malakhov, 1994), but also as a separate order (Triplonchida) (Siddiqi, 1983; Jairajpuri & Ahmad, 1992). Triplonchs can be easily distinguished from dorylaims by some very relevant morphological features, such as the complex nature of their protrusible structure, three-sectioned pharynx, arrangement of the pharyngeal glands, general morphology of the genital system and accessory elements, and absence of a prerectum (Decraemer, 1995). Molecular data have corroborated the position of triplonchs as a separate order of Enoplia.

Updated inventory of dorylaimid taxa

As will be clear from the above discussion, the taxonomy of the Dorylaimida has undergone significant changes since the order was first proposed, in particular during the 'golden age', when a myriad of new taxa were discovered and described. Leaving aside the two suborders, both monophyletic taxa, and the superfamilies, mostly non-monophyletic groups, the inventory of dorylaimid taxa currently includes 18 families, 363 nominal (290 valid) genera and 3410 nominal (3029 valid, 245 synonymous, 136 *incertae sedis* or *inquirendae*) species. An outline classification of subordinal, family, subfamily and generic rank taxa is presented in Table 14.3, which includes valid taxa and their accepted and/or tentative synonyms. This classification is an updated compilation from the available literature rather than a new proposal, a matter that would require much further analyses and which is outside the scope of this book.

Table 14.3. *Outline classification of Dorylaimida.*

SUBORDER
  FAMILY
    Subfamily
      Genus
NYGOLAIMINA Ahmad & Jairajpuri, 1979
  AETHOLAIMIDAE Jairajpuri, 1965
    Aetholaiminae Jairajpuri, 1965
      Aetholaimus Williams, 1962
  NYGELLIDAE Andrássy, 1958
    Nygellinae Andrássy, 1958
      Nygellus Thorne, 1939
  NYGOLAIMELLIDAE Clark, 1961
    Nygolaimellinae Clark, 1961
      Nygolaimellus Loos, 1949
  NYGOLAIMIDAE Thorne, 1935
    Nygolaiminae Thorne, 1935
      Afronygus Heyns, 1968
        = *Nygolaimus* (*Afronygus* Heyns, 1968)
      Aquatides Heyns, 1968
        = *Nygolaimus* (*Aquatides* Heyns, 1968)
      Clavicauda Heyns, 1968
        = *Nygolaimus* (*Clavicauda* Heyns, 1968)
      Clavicaudoides Heyns, 1968
        = *Nygolaimus* (*Clavicaudoides* Heyns, 1968)
      Feroxides Heyns, 1968
        = *Nygolaimus* (*Feroxides* Heyns, 1968)
      Laevides Heyns, 1968
        = *Nygolaimus* (*Laevides* Heyns, 1968)
      Nygolaimus Cobb, 1913
        = *Nygolaimus* (*Nygolaimus* Cobb, 1913)

Table 14.3. *(Continued.)*

Nygostylus Sukul, 1973, *inquirendum*
Paranygolaimus Heyns, 1968
= *Nygolaimus* (*Paranygolaimus* Heyns, 1968)
Pronygolaimus Sharma & Baqri, 2006
Solididens Heyns, 1968
= *Nygolaimus* (*Solididens* Heyns, 1968)
Paravulvinae Ahmad & Jairajpuri, 1982
Paravulvus Heyns, 1968
= *Nygolaimus* (*Paravulvus* Heyns, 1968)
DORYLAIMINA Pearse, 1936
ACTINOLAIMIDAE Thorne, 1939
= Brittonematidae Thorne, 1967
= Neoactinolaimidae Thorne, 1967
= Paractinolaimidae Thorne, 1967
Actinolaiminae Thorne, 1939
= Paractinolaiminae Thorne, 1967
Actinolaimus Cobb, 1913
Afractinolaimus Andrássy, 1970
Egtitus Thorne, 1967
Mactinolaimus Andrássy, 1970
Metactinolaimus Meyl, 1957
Neoactinolaimus Thorne, 1967
Nothactinolaimus Loof, 1973
Paractinolaimoides Khan, Ahmad & Jairajpuri, 1994
Paractinolaimus Meyl, 1957
= *Hexactinolaimus* Yeates, 1973
Scleroactinolaimus Ahmad, Khan & Ahmad, 1992
Stopractinca Khan, Ahmad & Jairajpuri, 1994
Brittonematinae Thorne, 1967
Actinca Andrássy, 1964
= *Actinca* (*Actinca* Andrássy, 1964)
Afractinca Vinciguerra & Clausi, 2000
Brasilaimus Lordello & Zamith, 1957
Brittonema Thorne, 1967
Parastomachoglossa Coomans & Loof, 1986
Practinocephalus Andrássy, 1974
= *Actinocephalus* Thorne, 1967, *nec* Stein (1848)
Westindicus Thorne, 1967
Trachypleurosinae Thorne, 1967
Trachactinolaimus Andrássy, 1963

Table 14.3. *(Continued.)*

Trachypleurosum Andrássy, 1959
= *Trachypleura* Thorne, 1939, *nec* Jackel (1900)
= *Dominiactinolaimus* Jairajpuri & Ahmad, 1992
APORCELAIMIDAE Heyns, 1965
Aporcelaiminae Heyns, 1965
Akrotonus Thorne, 1974
Aporcelaimellus Heyns, 1965
= *Takamangai* Yeates, 1967
Aporcelaimus Thorne & Swanger, 1936
Aporcelinus Andrássy, 2009
Aporcella Andrássy, 2002
Epacrolaimus Andrássy, 2000
Makatinus Heyns, 1965
Metaporcelaimus Lordello, 1965
= *Aporcelaimium* Loof & Coomans, 1970
Silvallis Ahmad & Jairajpuri, 1986
Tubixaba Monteiro & Lordello, 1980
Aporcedorinae Andrássy, 2009
Aporcedorus Jairajpuri & Ahmad, 1983
Sectonematinae Siddiqi, 1969
Aporcelaimoides Heyns, 1965
Scapidens Heyns, 1965
Sectonema Thorne, 1930
AULOLAIMOIDIDAE Jairajpuri, 1964
Adenolaimus Andrássy, 1973
Aulolaimoides Micoletzkty, 1915
Cladocephalus Swart & Heyns, 1991
Oostenbrinkia Ali, Suryawanshi & Ahmad, 1973
BELONDIRIDAE Thorne, 1939
= Axonchiidae Thorne, 1964
= Dorylaimellidae Jairajpuri, 1964
= Falcihastidae Siddiqi, 1968
= Oxydiridae Thorne, 1964
= Roqueidae Thorne, 1964
= Swangeriidae Jairajpuri, 1964
Belondirinae Thorne, 1939
= Axonchiinae Thorne, 1964
Amphibelondira Rahman, Jairajpuri, Ahmad & Ahmad, 1987
Anchobelondira Coomans & Nair, 1971
Axonchium Cobb, 1920

Table 14.3. *(Continued.)*

= *Dorylaimus* (*Axonchium* Cobb, 1920)
= *Axonchium* (*Axonchium* Cobb, 1920)
= *Axonchium* (*Hypaxonchium* Coomans & Nair, 1975)
= *Axonchium* (*Poraxonchium* Coomans & Nair, 1975)
Axonchoides Thorne, 1967
Belaxellus Thorne, 1974
Belondira Thorne, 1939
= *Belondirium* Andrássy, 1976
Belondirella Thorne, 1964
= *Yunqueus* Thorne, 1964
Bullaenema Sauer, 1968
Dactyluraxonchium Coomans & Nair, 1975
= *Axonchium* (*Dactyluraxonchium* Coomans & Nair, 1975)
Helicobelondira Yeates, 1973
Heynsaxonchium Coomans & Nair, 1975
= *Axonchium* (*Heynsaxonchium* Coomans & Nair, 1975)
Immanigula Andrássy, 1991
Metaxonchium Coomans & Nair, 1975
= *Axonchium* (*Metaxonchium* Coomans & Nair, 1975)
= *Axonchium* (*Discaxonchium* Coomans & Nair, 1975)
= *Axonchium* (*Epaxonchium* Coomans & Nair, 1975)
= *Axonchium* (*Spiculaxonchium* Ahmad & Jairajpuri, 1982)
Nimigula Andrássy, 1985
Phallaxonchium Jairajpuri & Dhanachand, 1979
Porternema Suryawanshi, 1972
= *Yubeldus* Khan, Azmi & Chawla, 1979
Probelondira Andrássy, 2009
Syncheilaxonchium Coomans & Nair, 1975
= *Axonchium* (*Syncheilaxonchium* Coomans & Nair, 1975)
Uniqaxonchium Dhanam & Jairajpuri, 1988
Dorylaimellinae Jairajpuri, 1964
Dorylaimellus Cobb, 1913
= *Dorylaimus* (*Dorylaimellus* Micoletzky, 1922)
= *Dorylaimellus* (*Nygellolaimellus* Dhanam & Jairajpuri, 1999)
= *Dorylaimellus* (*Prodorylaimellus* Jairajpuri & Ahmad, 1980)
= *Dorylaimus* (*Pungentoides* Altherr, 1950)
= *Amazonema* Siddiqi, 1983
= *Axodorylaimellus* Jairajpuri & Ahmad, 1980
= *Belondorylaimellus* Jairajpuri & Ahmad, 1980
= *Capitellus* Siddiqi, 1983
= *Elongidorylaimellus* Jairajpuri & Ahmad, 1980

Table 14.3. *(Continued.)*

= *Dorylaimellus* (*Filidorylaimellus* Jairajpuri & Ahmad, 1980)
= *Dorylaimellus* (*Clavidorylaimellus* Jairajpuri & Ahmad, 1980)
= *Ibadanus* Siddiqi, 1983
= *Jamilius* Siddiqi, 1983
= *Mesodorylaimellus* Jairajpuri & Ahmad, 1980
= *Metadorylaimellus* Jairajpuri & Ahmad, 1980
= *Dorylaimellus* (*Metadorylaimellus* Jairajpuri & Ahmad, 1980)
= *Dorylaimellus* (*Prodorylaimellus* Jairajpuri & Ahmad, 1980)
= *Rashidanema* Siddiqi, 1983
= *Sindellus* Siddiqi, 1983

Swangeriinae[1] Jairajpuri, 1964
= Falcihastinae Andrássy, 1976
= Hulquinae Siddiqi, 1982
= Oxydirinae Thorne, 1964
= Roqueinae Thorne, 1964
Durinemella Andrássy, 2009
= *Duriella* Dhanam & Jairajpuri, 1999, *nec* Row (1911)
Falcihasta Clark, 1964
Hulqus Siddiqi, 1982
Lindseyus Ferris & Ferris, 1973
Oxybelondira Ahmad & Jairajpuri, 1979
Oxydirus Thorne, 1939
Paraoxybelondira Dhanam & Jairajpuri, 1999
Paraoxydirus Jairajpuri & Ahmad, 1979
= *Tarjanius* Ferris, Goseco & Ferris, 1980
Paraqudsiella Siddiqi, 1982
Qudsiella Jairajpuri, 1967
Roqueus Thorne, 1964
Swangeria Thorne, 1939

DORYLAIMIDAE de Man, 1876
= Prodorylaimidae Andrássy, 1969

Afrodorylaiminae Andrássy, 1969
Afrodorylaimus Andrássy, 1964

Dorylaiminae de Man, 1876
= Arctidorylaiminae Mulvey & Anderson, 1979
= Laimydorinae Andrássy, 1969
= Mesodorylaiminae Andrássy, 1969
Baladorylaimus Andrássy, 2001
Calcaridorylaimus Andrássy, 1986
Calodorylaimus Andrássy, 1969

Table 14.3. *(Continued.)*

Chrysodorus Jiménez-Guirado & Cadenas, 1985
Crocodorylaimus Andrássy, 1988
Dorylaimus Dujardin, 1845
= *Dorylaimus* (*Dorylaimus* Dujardin, 1845)
Drepanodorylaimus Jairajpuri, 1966
Fuscheila Siddiqi, 1982
Halodorylaimus Andrássy, 1988
Idiodorylaimus Andrássy, 1969
Ischiodorylaimus Andrássy, 1969
Karadorylaimus Andrássy, 2011
Kittydorylaimus Andrássy, 1998
Laimydorus Siddiqi, 1969
Mesodorylaimus Andrássy, 1959
Miodorylaimus Andrássy, 1986
Miranema Thorne, 1939
Namaquanema Heyns & Swart, 1993
Paradorylaimus Andrássy, 1969
Rhinodorylaimus Ahmad, Baniyamuddin & Tauheed, 2010
Sclerodorylaimus Ahmad, Tauheed & Baniyamuddin, 2010
Labronematinae Peña-Santiago & Álvarez-Ortega, 2014
Crassogula Andrássy, 1991
Crassolabium Yeates 1967
= *Thonus* Thorne 1974
Heynsnema Peña-Santiago, Guerrero & Ciobanu, 2008
Labronema Thorne, 1939
= *Witoldinema* Brzeski, 1960
Labronemella Andrássy 1985
Nevadanema Álvarez-Ortega & Peña-Santiago 2012
Scalpelus Ahmad 2004
Skibbenema Van Reenen & Heyns 1986
Torumanawa Yeates 1967
Prodorylaiminae Andrássy, 1969
= Amphidorylaiminae Andrássy, 1976
Amphidorylaimus Andrássy, 1960
= *Minidorylaimus* Andrássy, 1972
Apodorylaimus Andrássy, 1988
Kunjudorylaimus Dhanam & Jairajpuri, 2000
Mitoaxonchium Yeates, 1973
Oxydiroides Altherr, 1972
Prodorylaimium Andrássy, 1969

Table 14.3. *(Continued.)*

Prodorylaimus Andrássy, 1959
Protodorylaimus Andrássy, 1988
Rafflesius Ahmad, 2007
LEPTONCHIDAE Thorne, 1935
= Basirotyleptidae Siddiqi & Khan, 1964
= Belonenchidae Thorne, 1964
= Tyleptidae Monteiro, 1970
Arganematinae Siddiqi, 2016
Arganema Siddiqi, 2016
Belonenchinae Thorne, 1964
Aculonchus Siddiqi, 1983
= *Basirotyleptus* (*Aculonchus* Siddiqi, 1983)
= *Basirotyleptus* (*Opisthotyleptus* Jairajpuri & Ahmad, 1992)
Basirotyleptus Jairajpuri, 1964
= *Basirotyleptus* (*Basirotyleptus* Jairajpuri, 1964)
= *Belonenchus* Thorne, 1964
= *Coronatyleptus* Siddiqi, 1982
= *Basirotyleptus* (*Coronatyleptus* Siddiqi, 1982)
= *Poncenema* Thorne, 1964
= *Punctoleptus* Khan, 1987
Glochidorella Siddiqi, 1982
Sclerostylus Goseco, Ferris & Ferris, 1981
Trichonchium Siddiqi & Khan, 1964
= *Basirotyleptus* (*Trichonchium* Siddiqi & Khan, 1964)
Zetalaimus Siddiqi, 1983
Leptonchinae Thorne, 1935
= Doryschotinae Khan, 1974
Apoleptonchus Siddiqi, 1982
Bertzuckermania Khera, 1970
Caveonchus Siddiqi, 1982
Clavigula Siddiqi, 1995
Funaria Van der Linde, 1938
= *Doryschota* Thorne, 1964
Incanema Andrássy, 1997
Lawtonema Siddiqi, 1999
Leptonchus Cobb, 1920
= *Neoleptonchus* Khan, 1974
Meylis Goseco, Ferris & Ferris, 1974
Paraleptonchus Dhanam & Jairajpuri, 1999
Proleptonchoides Ferris, Goseco & Kumar, 1979

Table 14.3. *(Continued.)*

Proleptonchus Lordello, 1955
= *Amphorostoma* Clark, 1962
= *Didoryllium* Izatullaeva, 1967
= *Picarilaimus* Sukul, 1973
Scalpenchinae Peña-Santiago, 2006
Scalpenchus Siddiqi, 1995
Tyleptinae Jairajpuri, 1964
Gymnotyleptus Ahmad & Jairajpuri, 1982
Tyleptus Thorne, 1939
Utahnema Thorne, 1939
= *Siddiqius* Andrássy, 1976
Xiphinemellinae Jairajpuri, 1964
Kantbhala Siddiqi, 1982
= *Longidorella* (*Kantbhala* Siddiqi, 1982)
Xiphinemella Loos, 1950
= *Taprobanus* Loos, 1949, *nec* Distant (1911)
= *Botalium* Heyns, 1963
= *Zalophidera* Siddiqi, 1982
LONGIDORIDAE[2] Thorne, 1935
= Xiphidoridae Khan, Chawla & Saha, 1978
= Xiphinematidae Dalmasso, 1969
Longidorinae Thorne, 1935
= Paralongidorinae Khan, Chawla & Saha, 1975
Longidoroides Khan, Chawla & Saha, 1978
Longidorus Micoletzky, 1922
= *Dorylaimus* (*Longidorus* Micoletzky, 1922)
= *Brevinema* Stegarescu, 1980
= *Neolongidorus* Khan, 1987
Paralongidorus Siddiqi, Hooper & Khan, 1963
= *Paralongidorus* (*Paralongidorus* Siddiqi, Hooper & Khan, 1963)
Siddiqia Khan, Chawla & Saha, 1978
= *Inagreius* Khan, 1982
= *Paralongidorus* (*Siddiqia* Khan, Chawla & Saha, 1978)
Xiphidorinae Khan, Chawla & Saha, 1978
Australodorus Coomans, Olmos, Casella & Chaves, 2004
Paraxiphidorus Coomans & Chaves, 1995
Xiphidorus Monteiro, 1976
Xiphinematinae Dalmasso, 1969
Xiphinema Cobb, 1913
= *Xiphinema* (*Basiphinema* Cohn & Sher, 1972)

Table 14.3. *(Continued.)*

= *Xiphinema* (*Diversiphinema* Cohn & Sher, 1972)
= *Xiphinema* (*Elongiphinema* Cohn & Sher, 1972)
= *Xiphinema* (*Halliphinema* Cohn & Sher, 1972)
= *Xiphinema* (*Krugiphinema* Cohn & Sher, 1972)
= *Xiphinema* (*Radiphinema* Cohn & Sher, 1972)
= *Xiphinema* (*Rotundiphinema* Cohn & Sher, 1972)
= *Xiphinema* (*Xiphinema* Cobb, 1913)

MYDONOMIDAE Thorne, 1964
= Dorylaimoididae Siddiqi, 1969
= Miranematidae Siddiqi, 1969

Calolaiminae Goseco, Ferris & Ferris, 1976
Calolaimus Timm, 1964
= *Galophinema* Siddiqi, 1965
Paratimmus Baniyamuddin & Ahmad, 2009
Timmus Goseco, Ferris & Ferris, 1976

Mydonominae Thorne, 1964
= Dorylaimoidinae Siddiqi, 1969
Dorylaimoides Thorne & Swanger, 1936
= *Tarjania* Brzeski & Szczygiel, 1961
= *Dorylaimoides* (*Tarjania* Brzeski & Szczygiel, 1961)
= *Leptonema* Jairajpuri, 1961
= *Shamimonema* Chawla, Khan & Prasad, 1965
= *Dorylaimoides* (*Arcidorylaimoides* Jairajpuri & Ahmad, 1992)
= *Dorylaimoides* (*Digidorylaimoides* Jairajpuri & Ahmad, 1992)
= *Dorylaimoides* (*Longidorylaimoides* Jairajpuri & Ahmad, 1992)
Morasia Baqri & Jairajpuri, 1969
Mydonomus Thorne, 1964

NORDIIDAE Jairajpuri & Siddiqi, 1964
= Encholaimidae Golden & Murphy, 1967
= Kochinematidae Darekar & Khan, 1979
= Pararhyssocolpidae Elshishka, Lazarova, Radoslavov, Hristov & Peneva, 2015
= Pungentidae Monteiro, 1970

Actinolaimoidinae Jairajpuri & Ahmad, 1992
Actinolaimoides Meyl, 1957
= *Enchodelium* Andrássy, 1963
= *Longidorella* (*Actinolaimoides* Meyl, 1957)

Oriverutinae Andrássy, 2009
Malekus Thorne, 1974
Oonaguntus Thorne, 1974

Table 14.3. *(Continued.)*

Oriverutoides Ahmad & Sturhan, 2002
Oriverutus Siddiqi, 1971
= *Inbionema* Loof & Zullini, 2000
= *Mammillonema* Darekar & Khan, 1982
= *Ecanema* Ahmad & Shaheen, 2005
Paroriverutus Carbonell & Coomans, 1982
Stomacholaimus Andrássy, 2011
Encholaiminae Golden & Murphy, 1967
= Cephalodorylaiminae Jairajpuri, 1967
= Helmabiinae Siddiqi, 1971
Acephalodorylaimus Ahmad & Jairajpuri, 1983
Cephalodorylaimus Jairajpuri, 1967
Echinodorus Siddiqi, 1995
Encholaimus Golden & Murphy, 1967
Helmabia Siddiqi, 1971
Nemabia Siddiqi, 1995
Nordiinae Jairajpuri & Siddiqi, 1964
= Enchodorellinae Andrássy, 1966
Acunemella Andrássy, 2002
= *Longidorella* (*Acunemella* Andrássy, 2002)
Longidorella Thorne, 1939
= *Longidorella* (*Longidorella* Thorne, 1939)
= *Enchodorella* Khan, 1964
= *Longidorella* (*Enchodorella* Khan, 1964)
= *Nordia* Jairajpuri & Siddiqi, 1964
= *Saevadorella* Siddiqi, 1982
= *Longidorella* (*Saevadorella* Siddiqi, 1982)
Thornedia Husain & Khan, 1965
= *Longidorella* (*Thornedia* Husain & Khan, 1965)
Pungentinae Siddiqi, 1969
= Californidorinae Robbins & Weiner, 1978
= Lenonchiinae Eliava, 1978
Californidorus Robbins & Weiner, 1978
Enchodeloides Elshishka, Lazarova, Radoslavov, Hristov, & Peneva, 2017
Enchodelus Thorne, 1939
= *Enchodelus* (*Enchodelus* Thorne, 1939)
= *Enchodelus* (*Rotundus* Ahmad & Jairajpuri, 1980)
Enchodorus Vinciguerra, 1976
Heterodorus Altherr, 1952

Table 14.3. *(Continued.)*

= *Enchodelus* (*Heterodorus* Altherr, 1952)
= *Enchodelus* (*Nepalus* Ahmad & Jairajpuri, 1980)
= *Enchodelus* (*Paraenchodelus* Ahmad & Jairajpuri, 1980)
Kochinema Heyns, 1963
Lanzavecchia Zullini, 1988
Lenonchium Siddiqi, 1965
Papuadorus Andrássy, 2009
Pararhyssocolpus Elshishka, Lazarova, Radoslavov, Hristov & Peneva, 2015
Pungentella Andrássy, 2009
Pungentus Thorne & Swanger, 1936
= *Paramonovia* Eliava, 1973
Rhyssocolpus Andrássy, 1971
Stenodorylaimus Álvarez-Ortega & Peña-Santiago, 2011
QUDSIANEMATIDAE Jairajpuri, 1965
= Chrysonematidae Siddiqi, 1969
= Crateronematidae Siddiqi, 1969
= Discolaimidae Siddiqi, 1969
= Paraxonchiidae Dhanachand & Jairajpuri, 1981
Arctidorylaiminae Mulvey & Anderson, 1979
Arctidorylaimus Mulvey & Anderson, 1979
Carcharolaiminae Thorne, 1967
Antholaimus Cobb, 1913
Carcharolaimus Thorne, 1939
= *Carcharodiscus* Andrássy, 1992
Caribenema Thorne, 1967
Caryboca Lordello, 1967
= *Carcharoides* Thorne, 1967
Crateronematinae Siddiqi, 1969
Chrysonema Thorne, 1929
= *Chrysonemoides* Siddiqi, 1969
Crateronema Siddiqi, 1969
Discolaiminae Siddiqi, 1969
Discolaimium Thorne, 1939
= *Durinema* Jairajpuri, 1968
= *Neodiscolaimium* Patil & Khan, 1982
Discolaimoides Heyns, 1963
Discolaimus Cobb, 1913
= *Dorylaimus* (*Discolaimus* Cobb, 1913)
= *Ottolaimus* Kirjanova, 1951

Table 14.3. *(Continued.)*

Filidiscolaimus Siddiqi, 1995
Latocephalus Patil & Khan, 1982
Mylodiscoides Lordello, 1963
Mylodiscus Thorne, 1939
Salimella Siddiqi, 2005
Lordellonematinae Siddiqi, 1969
Lordellonema Andrássy, 1960
= *Poronema* Heyns, 1963
Moshajia Siddiqi, 1982
Poronemella Siddiqi, 1969
Sicorinema Siddiqi, 1982
Sicorinemella Andrássy, 2009
Pachydorylaiminae Andrássy, 2009
Pachydorylaimus Siddiqi, 1983
Paraxonchiinae Dhanachand & Jairajpuri, 1981
= Gopalinae Andrássy, 2009
Cerviconema Andrássy, 2009
Gopalus Khan, Jairajpuri & Ahmad, 1988
Parapalus Loof & Zullini, 2000
Paraxonchium Krall, 1958
= *Drepanodorus* Altherr, 1954, *nec* Menge (1869)
= *Drepanophorus* Altherr, 1968
Ramphidorylaimus Baniyamuddin, Ahmad & Jairajpuri, 2010
Tendinema Siddiqi, 1995
Qudsianematinae Jairajpuri, 1965
= Ecumenicinae Andrássy, 2009
= Eudorylaiminae Khan & Fatima, 1980
Allodorylaimus Andrássy, 1986
Amblydorylaimus Andrássy, 1998
Baqriella Ahmad & Jairajpuri, 1989
Boreolaimus Andrássy, 1988
Cricodorylaimus Ahmad & Sturhan, 2001
Dorydorella Andrássy, 1987
Ecumenicus Thorne, 1974
= *Indokochinema* Darekar & Khan, 1979
Epidorylaimus Andrássy, 1986
Eudorylaimus Andrássy, 1959
= *Qudsianema* Jairajpuri, 1965
Kallidorylaimus Andrássy, 1989
Kolodorylaimus Andrássy, 1998

Table 14.3. *(Continued.)*

Microdorylaimus Andrássy, 1986
Talanema Andrássy, 1991
THORNIIDAE De Coninck, 1965
Thorniinae De Coninck, 1965
Nygolaimoides Meyl *in* Andrássy, 1960
Thornia Meyl, 1954
Thorniosa Andrássy, 1996
Thorneellinae Andrássy, 1987
Loofilaimus Jairajpuri, Ahmad & Sturhan, 1998
Sphaeroamphis Ahmad & Sturhan, 2000
Thorneella Andrássy, 1960
THORNENEMATIDAE Siddiqi, 1969
Thornenematinae Siddiqi, 1969
= Medalinematinae Baqri & Jana, 1980
Coomansinema Ahmad & Jairajpuri, 1969
Coomansinemoides Sen, Chatterjee & Manna, 2012
Indodorylaimus Ali & Prabha, 1974
Lagenonema Andrássy, 1987
Laurophragus Nesterov, 1976
Prothornenema Baqri & Bohra, 2003
Opisthodorylaimus Ahmad & Jairajpuri, 1982
Sicaguttur Siddiqi, 1971
= *Mehdinema* Baqri & Jana, 1980
= *Medalinema* Baqri & Jana, 1980
Thornenema Andrássy, 1959
= *Jairajpuria* Baqri & Jana, 1980
Willinematinae Andrássy, 1987
Anadorella Siddiqi, 2005
Paratimminema Rahaman, Ahmad & Khan, 1993
Sclerolabia Carbonell & Coomans, 1986
Timminema Khan, 1978
Willinema Baqri & Jairajpuri, 1967
TYLENCHOLAIMELLIDAE Jairajpuri, 1964
Athernematinae Ahmad & Jairajpuri, 1978
Athernema Ahmad & Jairajpuri, 1978
Tylenchollaimellinae Jairajpuri, 1964
= Dorylliinae Izatullaeva, 1967
Agmodorus Thorne, 1964
Dorella Jairajpuri, 1964
Doryllium Cobb, 1920

Table 14.3. *(Continued.)*

= *Dorylaimus* (*Doryllium* Micoletzky, 1922)
= *Gerthus* Goseco, Ferris & Ferris, 1975
Goferus Jairajpuri & Ahmad, 1992
Margollus Peña-Santiago, Peralta & Siddiqi, 1993
Oostenbrinkella Jairajpuri, 1965
Phellonema Thorne, 1964
Tylencholaimellus Cobb *in* Cobb, 1915
= *Tylencholaimellus* (*Coronatylencholaimellus* Jairajpuri & Ahmad, 1992)
= *Tylencholaimellus* (*Tylencholaimellus* Cobb *in* Cobb, 1915)
TYLENCHOLAIMIDAE Filipjev, 1934
Metadorylaiminae Andrássy, 1976
Metadorylaimus Jairajpuri & Goodey, 1966
Neometadorylaimus Jairajpuri & Ahmad, 1992
Tylencholaiminae Filipjev, 1934
= Discomyctinae Andrássy, 2009
= Mumtaziinae Andrássy, 1976
= Paratylencholaiminae Wu, Xu, Xie & Wang, 2019
Capilonchus Siddiqi, 1982
Chitwoodielloides Ahmad & Araki, 2003
Chitwoodiellus Jiménez-Guirado & Peña-Santiago, 1992
Chitwoodius Furstenberg & Heyns, 1966
= *Chitwoodia* Furstenberg & Heyns, 1966, *nec* Gerlach (1956)
Discomyctus Thorne, 1939
Loncharionema Goseco, Ferris & Ferris, 1974
Meylonema Andrássy, 1960
= *Leptonchulus* Chawla, Bhamburkar, Khan & Prasad, 1968
Mumtazium Siddiqi, 1969
Promumtazium Siddiqi, 1982
= *Tsukubanema* Ahmad & Araki, 2002
Pseudotylencholaimus Jairajpuri & Ahmad, 1992
Rostrulium Siddiqi, 1995
Sclerolaimus Jairajpuri & Ahmad, 1992
Tantunema Siddiqi, 1982
Tylenchodoroides Gagarin, 2001
= *Tylenchodorus* Gagarin, 2000, *nec* Siddiqi (1983)
Tylenchodorus Siddiqi, 1983
Tylencholaimus de Man, 1876
= *Tylencholaimus* (*Amphitylencholaimus* Jairajpuri & Ahmad, 1992)
= *Tylencholaimus* (*Disctylencholaimus* Jairajpuri & Ahmad, 1992)

Table 14.3. *(Continued.)*

= *Tylencholaimus* (*Leptotylencholaimus* Jairajpuri & Ahmad, 1992)
= *Tylencholaimus* (*Opisthotylencholaimus* Jairajpuri & Ahmad, 1992)
= *Tylencholaimus* (*Protylencholaimus* Jairajpuri & Ahmad, 1992)
= *Tylencholaimus* (*Tylencholaimus* de Man, 1876)
= *Xenonchium* Siddiqi & Khan, 1964
= *Paratylencholaimus* Wu, Xu, Xie & Wang, 2019
Wasimellus Bloemers & Wanless, 1996
Vanderlindiinae Siddiqi, 1969
Curvidorylaimus Jairajpuri & Rahman, 1983
Vanderlindia Heyns, 1964
DORYGASTRIDAE Khera, 1970, *inquirendum*
Dorygaster Khera, 1970, *inquirendum*
Genus inquirendum and/or *incertae sedis*:
Opailaimus Kirjanova, 1951

[1] Coomans & Kheiri (1986) divided this subfamily into three tribes, namely Swangeriini, Falcihastini and Roqueini.
[2] Coomans (1985) proposed a somewhat different scheme for the classification of Longidoridae, with two subfamilies, Xiphinematinae and Longidorinae, the latter consisting of two tribes, Xiphidorini and Longidorini.

References

ANDRÁSSY, I. (1976). *Evolution as a basis for the systematization of nematodes*. London, UK, Pitman Publishing Ltd.

ANDRÁSSY, I. (2009). *Free-living nematodes of Hungary, III (Nematoda errantia)*. In: *Pedozoologica Hungarica 5* (Series Editors: Csuzdi, C & Mahunka, S.). Budapest, Hungary, Hungarian Natural History Museum and Systematic Zoology Research Group of the Hungarian Academy of Sciences.

BASTIAN, H.C. (1865). Monograph of the Anguillulidae, or free nematoids, marine, land and freshwater; with descriptions of 100 new species. *Transactions of the Linnean Society of London, Zoology* 25, 73-184.

BERT, W., KARSSEN, G. & HELDER, J. (2011). Phylogeny and evolution of nematodes. In: Jones, J., Gheysen, G. & Fenoll, C. (Eds). *Genomics and molecular genetics of plant-nematode interactions*. Dordrecht, The Netherlands, Springer Science, pp. 45-59.

BLAXTER, M.L., DE LEY, P., GAREY, J.R., LIU, L.X., SCHELDEMAN, P., VIERSTRAETE, A., VANFLATEREN, J.R., MACKEY, L.Y., DORRIS, M., FRISSE, L.M. *et al.* (1998). A molecular evolutionary framework for the phylum Nematoda. *Nature* 392, 71-75. DOI: 10.1038/32160

BÜTSCHLI, O. (1873). Beiträge zur Kenntnis der freilebenden Nematoden. *Nova Acta der Kaiserlich Leopoldinisch-Carolinischen Deutschen Akademie der Natutforscher* 36, 1-144.

CLARK, W.C. (1961). A revised classification of the order Enoplida (Nematoda). *New Zealand Journal of Science* 4, 123-150.

COBB, N.A. (1893). Nematodes mostly Australian and Fijian. *Macleay Memorial Volume, Linnean Society of New South Wales* 13, 1-59.

COBB, N.A. (1920). One hundred new nemas. (Type species of 100 new genera). *Contributions to a science of nematology* 9, 217-243.

CARBONELL, E. & COOMANS, A. (1987). The genus *Thornenema* Andrássy, 1959 (Nematoda: Thornenematidae). *Nematologica* 32(1986), 125-179. DOI: 10.1163/187529286X00129

COOMANS, A. (1985). A phylogenetic approach to the classification of the Longidoridae (Nematoda: Dorylaimida). *Agriculture, Ecosystem and Environment* 12, 335-354. DOI: 10.1016/0167-8809(85)90006-4

COOMANS, A. & KHEIRI, A. (1986). Observations on *Lindseyus costatus* Ferris & Ferris, 1973 with a discussion on its relationships (Nematoda: Belondiridae). *Revue de Nématologie* 9, 357-367.

COOMANS, A. & LOOF, P.A.A. (1970). Morphology and taxonomy of Bathyodontina (Dorylaimida). *Nematologica* 16, 180-196. DOI: 10.1163/187529270X00199

COOMANS, A. & LOOF, P.A.A. (1986). Observations on the glands of the male reproductive system in dorylaims and its phylogenetic importance. *Revue de Nématologie* 9, 261-265.

COOMANS, A. & NAIR, P. (1975). The genus *Axonchium* (Nematoda: Belondiridae). VI. Atypical species, keys, subgenera and conclusions. *Nematologica* 21, 296-332. DOI: 10.1163/187529275X00059

COOMANS, A. & VAN DER HEIDEN, A. (1979). The systematic position of the family Ironidae and its relation to the Dorylaimida. *Annales de la Société Royal Zoologique de Belgique* 10, 5-11.

DE CONINCK, L. (1965). Systématique des nématodes. In: Grassé, P.P. (Ed.). *Traité de zoologie – anatomie, systématique, biologie. Némathelmitnthes (Nématodes). Tome IV, fascicule II.* Paris, France, Masson, pp. 586-731.

DECRAEMER, W. (1995). *The family Trichodoridae: stubby root and virus vector nematodes.* Dordrecht, The Netherlands, Kluwer Academic Publishers.

DE LEY, P. & BLAXTER, M. (2002). Systematic position and phylogeny. In: Lee, D.L. (Ed.). *The biology of nematodes.* London, UK, Taylor & Francis, pp. 1-30.

DE LEY, P. & BLAXTER, M. (2004). A new system for Nematoda: combining morphological characters with molecular trees, and translating clades into ranks and taxa. In: Cook, R. & Hunt, D.J. (Eds). *Nematology Monographs and Perspectives 2* (Series Editors: Cook, R. & Hunt, D.J.). Leiden, The Netherlands, Brill, pp. 633-653.

DE MAN, J.G. (1876). Onderzoekingen over vrij in de aarde levende Nematoden. *Tijdschrift der Nederlandsche Dierkundige Vereeniging* 2, 78-196.

DE MAN, J.G. (1880). Die einheimischen, frei in der reinen Erde und im süssen Wasser lebenden Nematoden. *Tijdschrift der Nederlandsche Dierkundige Vereeniging* 5, 1-104.

DUJARDIN, F. (1845). *Histoire naturelle des helminthes ou vers intestinaux.* Paris, France, Librairie Encyclopaedique de Roret.

FILIPJEV, I.N. (1929). The classification of free-living nematodes and relations to parasitic forms. *Journal of Parasitology* 15, 281.

FILIPJEV, I.N. (1934). The classification of free-living nematodes and their relation to parasitic nematodes. *Smithsonian Miscellaneous Collection* 89, 1-63.

GOODEY, J.B. (1963). *Soil and freshwater nematodes*, 2nd edition. London, UK, Methuen.

GOSECO, C.G., FERRIS, V.R. & FERRIS, J.M. (1974). Revisions in Leptonchoidea (Nematoda: Dorylaimida). *Leptonchus, Proleptonchus, Funaria*, and *Meylis* n. gen. in Leptonchidae, Leptonchinae. *Purdue University Agricultural Experiment Station Research Bulletin* 911, 1-32.

HEYNS, J. (1965). On the morphology and taxonomy of the Aporcelaimidae, a new family of dorylaimid nematodes. *Entomology Memoirs, Department of Agriculture Technical Services, Republic of South Africa* 10, 1-51.

HEYNS, J. (1968). A monographic study of the nematode families Nygolaimidae and Nygolaimellidae. *Entomology Memoirs, Plant Protection Research Institute Pretoria, South Africa* 19, 1-144.

HODDA, M. (2007). Phylum Nematoda. In: Zhang, Z.Q. & Shear, W.A. (Eds). Linnaeus tercentenary: progress in invertebrate taxonomy. *Zootaxa* 1668, 265-293.

HODDA, M. (2011). Phylum Nematoda Cobb, 1932. In: Z.-Q. Zhang (Ed.). *Animal biodiversity: An outline of higher-level classification and survey of taxonomic richness. Zootaxa Monograph 3148.* Auckland, New Zealand, Magnolia Press, pp. 63-95.

HOLOVACHOV, O. (2014). Order Isolaimida Cobb, 1920. In: Schmidt-Rhaesa, A. (Ed.). *Handbook of zoology – Gastrotricha, Cycloneuralia and Gnathifera. Volume 2. Nematoda.* Berlin, Germany, De Gruyter, pp. 313-316.

HOLTERMAN, M., WURFF, A. VAN DER, ELSEN, S. VAN DEN MEGEN, H., BONGERS, T., HOLOVACHOV, O., BAKKER, J. & HELDER, J. (2006). Phylum-wide analysis of SSU rDNA reveals deep phylogenetic relationships among nematodes and accelerated evolution toward crown clades. *Molecular Biology and Evolution* 23, 1792-1800. DOI: 10.1093/molbev/msl044.

HOLTERMAN, M., RYBARCZYK, K., VAN DEN ESSEN, S., VAN MEGEN, H., MOOYMAN, P., PEÑA-SANTIAGO, R., BONGERS, T., BAKKER, J. & HELDER, J. (2008). A ribosomal DNA-based framework for the detection and quantification of stress-sensitive nematode families in terrestrial habitats. *Molecular Ecology Resources* 8, 23-34. DOI: 10.1111/j.1471-8286.2007.01963.x

INGLIS, W.G. (1983). An outline classification of the phylum Nematoda. *Australian Journal of Zoology* 31, 243-255. DOI: 10.1071/zo9830243

JAIRAJPURI, M.S. (1969). Studies on Mononchida of India. I. The genera *Hadronchus*, *Iotonchus* and *Miconchus* and a revised classification of Mononchida, new order. *Nematologica* 15, 557-581. DOI: 10.1163/187529269X00894

JAIRAJPURI, M.S. (1983). Observations on *Campydora* (Nematoda: Dorylaimida). *Nematologia Mediterranea* 11, 33-42.

JAIRAJPURI, M.S. & AHMAD, W. (1992). *Dorylaimida. Free-living, predaceous and plant-parasitic nematodes*. Leiden, The Netherlands, Brill.

LORENZEN, S. (1994). *The phylogenetic systematics of free-living nematodes*. London, UK, The Ray Society.

MAGGENTI, A.R. (1982). Nemata. In: Parker, S.P. (Ed.). *Synopsis and classification of living organisms, Volume 1*. New York, NY, USA, McGraw-Hill, Inc., pp. 879-929.

MALAKHOV, V.V. (1994). *Nematodes: structure, development, classification and phylogeny*. Washington, USA, Smithsonian Institution Press.

MALAKHOV, V.V. (1998). Embryological and histological peculiarities of the order Enoplida, a primitive group of nematodes. *Russian Journal of Nematology* 6, 41-46.

MEDAL, B.H.M., DEBENHAM, N.J., DE LEY, P., DE LEY, I.T., VANFLETEREN, J.R., VIESTRAETE, A.R., BERT, W., BORGONIE, G., MOENS, T., TYLER, P.A. *et al.* (2007). An improved molecular phylogeny of the Nematoda with special emphasis on marine taxa. *Molecular Phylogenetics and Evolution* 42, 622-636. DOI: 10.1016/j.ympev.2006.08.025

MILJUTIN, D.M. (2014). Order Marimermithida Rubtzov 1980, emend. Tchesunov 1995. In: Schmidt-Rhaesa, A. (Ed.). *Handbook of zoology – Gastrotricha, Cycloneuralia and Gnathifera. Volume 2. Nematoda*. Berlin, Germany, De Gruyter, pp. 345-350.

MULLIN, P.G., HARRIS, T.S. & POWERS, T.O. (2004). Systematic status of *Campydora* Cobb, 1920 (Nematoda: Campydorina). *Nematology* 5(2003), 699-711. DOI: 10.1163/156854103322746878

MULLIN, P.G., HARRIS, T.S. & POWERS, T.O. (2005). Phylogenetic relationships of Nygolaimina and Dorylaimina (Nematoda: Dorylaimida) inferred from small subunit ribosomal DNA sequences. *Nematology* 7, 59-79. DOI: 10.1163/1568541054192199

ÖRLEY, L. (1880). *Monographie der Anguilluliden. Eine von der K. ung. Naturhistorischen Gesellschraft gekrönte Preisschrift*. Budapest, Hungary.

PEARSE, A.S. (1936). *Zoological names. A list of phyla, classes and orders prepared for Section F. American Association for the Advancement of Science*. Durhan, NC, USA, Duke University Press.

PEARSE, A.S. (1942). *Introduction to parasitology*. Baltimore, MD, USA, Springfield, ILL.

PEÑA-SANTIAGO, R. (2006). Dorylaimida Part I: Superfamilies Belondiroidea, Nygolaimoidea and Tylencholaimoidea. In: Eyualem-Abebe, Andrássy, I. & Trunspurger, W. (Eds). *Freshwater nematodes: ecology and taxonomy*, Wallingford, UK, CAB International, pp. 326-391.

PEÑA-SANTIAGO, R. (2013). *In memoriam* – Andrássy's legacy: a unique, essential and insuperable contribution to Nematology. *Journal of Nematode Morphology and Systematics* 16, 1-16.

PEÑA-SANTIAGO, R. (2014). Order Dorylaimida Pearse, 1942. In: Schmidt-Rhaesa, A. (Ed.). *Handbook of zoology – Gastrotricha, Cycloneuralia and Gnathifera. Volume 2. Nematoda*. Berlin, Germany, De Gruyter, pp. 277-297.

RUSIN, L.Y., ALEKSHIN, V.V., TCHESUNOV, A.V. & ATRASHKEVICH, G.I. (2003). The 18S ribosomal RNA gene of *Soboliphyme baturini* Petrow, 1930 (Nematoda: Dioctophymida) and its implications for phylogenetic relationships within Dorylaimia. *Nematology* 5, 615-628. DOI: 10.1163/156854103322683337

SIDDIQI, M.R. (1969). *Crateronema* n. gen. (Crateronematidae n. fam.), *Poronemella* n. gen. (Lordellonematinae n. subfam.) and *Chrysonemoides* n. gen. (Chrysonematidae n. fam.) with a revised classification of Dorylaimoidea (Nematoda). *Nematologica* 15, 81-100. DOI: 10.1163/187529269X00128

SIDDIQI, M.R. (1983). Phylogenetic relationships of the soil nematode orders Dorylaimida, Mononchida, Triplonchida and Alaimida, with a revised classification of the subclass Enoplia. *Pakistan Journal of Nematology* 1, 79-110.

SPRATT, D.M. (2014). Order Muspiceida. In: Schmidt-Rhaesa, A. (Ed.). *Handbook of zoology – Gastrotricha, Cycloneuralia and Gnathifera. Volume 2. Nematoda*. Berlin, Germany, De Gruyter, pp. 335-344.

SPRATT, D.M. & NICHOLAS, W.L. (2002). Morphological evidence for the systematic position of the order Muspiceida (Nematoda). *Transactions of the Royal Society of South Australia* 126, 51-62.

THORNE, G. (1934). The classification of the higher groups of dorylaims. *Proceedings of the Helminthological Society of Washington* 1, 19.

THORNE, G. (1935). Notes on free-living and plant-parasitic nematodes. II. Higher classification groups of Dorylaimoidea. *Proceedings of the Helminthological Society of Washington* 2, 96-98.

THORNE, G. (1939). A monograph of the nematodes of the superfamily Dorylaimoidea. *Capita Zoologica* 8, 1-261.

THORNE, G. (1964). Nematodes of Puerto Rico: Belondiroidea, new superfamily, Leptonchidae Thorne, 1935, and Belonenchidae new family (Nematoda: Adenophorea, Dorylaimida). *University of Puerto Rico Agriculture Experimental Station Technical Paper* 39, pp. 1-51.

THORNE, G. (1967). Nematodes of Puerto Rico: Actinolaimoidea, new superfamily with a revision of its genera and species with addenda to Belondiroidea (Nematoda, Adenophorea, Dorylaimida). *University of Puerto Rico Agriculture Experimental Station Technical Paper* 43, 1-48.

THORNE, G. & SWANGER, H.H. (1936). A monograph of the nematode genera *Dorylaimus* Dujardin, *Aporcelaimus* n. gen., *Dorylaimoides* n. gen., and *Pungentus* n. gen. *Capita Zoologica* 6, 1-223.

VAN MEGEN, H., VAN DEN ELSEN, S., HOLTERMAN, M., KARSSEN, G., MOOYMAN, P., BONGERS, T., HOLOVACHOV, O., BAKKER, J. & HELDER, J. (2009). A phylogenetic tree of nematodes based on about 1200 full-length small subunit ribosomal DNA sequences. *Nematology* 11, 927-950. DOI: 10.1163/156854109X456862

VINCIGUERRA, M.T. (1988). A new classification of Actinolaimoidea Thorne, 1939 using a cladistic approach. *Nematologica* 33, 251-277. DOI: 10.1163/187529287X00399

VINCIGUERRA, M.T. (2006). Dorylaimida Part II: Superfamily Dorylaimoidea. In: Eyualem-Abebe, Andrássy, I. & Traunspurger, W. (Eds). *Freshwater nematodes: ecology and taxonomy*. Wallingford, UK, CAB International, pp. 392-467.

YUSHIN, V.V. & MALAKHOV, V.V. (2004). Spermatogenesis and nematode phylogeny. In: Cook, R. & Hunt, D.J. (Eds). *Nematology Monographs and Perspectives 2* (Series Editors: Cook, R. & Hunt, D.J.). Leiden, The Netherlands, Brill, pp. 655-665.

Nematology Monographs & Perspectives, 2021, Vol. 13, 269-272

Subject Index

Cited Taxa Index

www.ingramcontent.com/pod-product-compliance
Lightning Source LLC
Chambersburg PA
CBHW071229230926
51849CB00001B/58
9789004439993